SYNTHETIC FRONTIERS

SYNTHETIC FRONTIERS

Ocean Plastic and the Persistence of Trash Islands

KIM DE WOLFF

The MIT Press
Cambridge, Massachusetts
London, England

The MIT Press
Massachusetts Institute of Technology
77 Massachusetts Avenue, Cambridge, MA 02139
mitpress.mit.edu

The MIT Press would like to thank the anonymous peer reviewers who provided comments on drafts of this book. The generous work of academic experts is essential for establishing the authority and quality of our publications. We acknowledge with gratitude the contributions of these otherwise uncredited readers.

This book was set in Adobe Garamond and Berthold Akzidenz Grotesk by Westchester Publishing Services. Printed and bound in the United States of America.

Library of Congress Cataloging-in-Publication Data

Names: De Wolff, Kim, 1981– author
Title: Synthetic frontiers : ocean plastic and the persistence of trash islands / Kim De Wolff.
Description: Cambridge, MA : The MIT Press, [2025] | Includes bibliographical references and index.
Identifiers: LCCN 2025016464 (print) | LCCN 2025016465 (ebook) | ISBN 9780262553681 paperback | ISBN 9780262385329 epub | ISBN 9780262385336 pdf
Subjects: LCSH: Plastic marine debris | Marine pollution—Prevention
Classification: LCC TD427.P62 .D49 2025 (print) | LCC TD427.P62 (ebook)
LC record available at https://lccn.loc.gov/2025016464
LC ebook record available at https://lccn.loc.gov/2025016465

EU Authorised Representative: Easy Access System Europe, Mustamäe tee 50, 10621 Tallinn, Estonia | Email: gpsr.requests@easproject.com

Contents

Acknowledgments *vii*

INTRODUCTION *1*

IN/OUT OF PLACE *31*

1 **PLASTIC COASTLINES, SYNTHETIC FRONTIERS** *35*

REORIENTATIONS *65*

2 **THE TRASH ISLAND THAT ISN'T THERE** *69*

ENTANGLEMENTS *97*

3 **LIVING IN THE PLASTISPHERE** *101*

LANDINGS *129*

SYNTHETICS: PLACING TOGETHER/WITH *133*

Notes *143*
Bibliography *173*
Index *191*

Acknowledgments

This book is made together/with the ongoing support and generosity of an abundance of kind, brilliant, caring humans. My own relations to peoples and places have been very much transformed in the years since I stepped aboard the *Sea Dragon*. A project that began in a graduate student office overlooking the Pacific Ocean in San Diego winds its way to conclusion atop the underground sprawl of fracking wells in North Texas. What once seemed like a strange irony—writing about the sea in the Cross Timbers and Blackland prairie—has challenged me to rethink what exactly constitutes the frontlines of an ocean plastic crisis. I continue to grapple with my thorny inheritances as a settler, moving from the unceded territory of the Kumeyaay Nation, to where I currently work and reside on the unceded lands of the Wichita and Caddo Affiliated tribes. In Hawai'i, where the descendants of the original people are Kānaka ʻōiwi, I am a visitor in an Indigenous space that remains illegally occupied by the United States. Such acknowledgments are not ends in themselves, but ways of situating ongoing responsibilities together/with land and each other that must extend beyond recognition. I am especially grateful to those who continue to model best practices. I continue to listen.

This project would not exist without the enthusiastic welcome from Algalita Marine Research and Education and their affiliates, whose company I have very much enjoyed revisiting in writing this book. My collaborators in Long Beach: Katie Allen, Marieta Francis, Jeanne Gallagher, Gwen Lattin, Charles Moore, and Ann Zellers. My shipmates in the close quarters of the *Sea Dragon:* Judy Bolquardsen, Carolynn Box, Hank Carsen, Clive

Cosby, Marcus Eriksen, Womchin Jin, Brandon Kim, Rob Johnson, Ming Hui Liao, Karen Ristuben, Dale Selvam, and Tim Silverwood. Among the vast community of concern for plastics and oceans, I give my thanks for sharing their time and expertise to Miriam Goldstein, Jan Hafner, Lindsey Hoshaw, Nikolai Maximenko, Yoshiko Ohkura, Noni Sanford, Catherine Spina, Hideshige Takada, Elizabeth Venrick, and Angelicque White.

Finishing this book been a gift of re-turning to formative academic experiences whose import I have only been able to grasp in full many years later. The University of California, San Diego Communication Department nurtured creative and strong foundations from which this project could grow. My extended gratitude to Chandra Mukerji, for showing me how to write theory as story, and who undoubtedly saw how this project was about territorial power all along. I owe my overdue thanks to Gary Fields and Val Hartouni, whose graduate courses have had so much more to do with my work than I understood at the time. As a postdoctoral researcher, the University of California Merced Center for the Humanities provided a home where the hydrohumanities could flourish. For ushering in my water era, I would like to especially thank Ruth Mostern, Rina Faletti, Ignacio López-Calvo, and Kevin Dawson; along with the Meander Collective: Irene Klaver, Elana Zilberg, Stephanie Kane, and Matilde Córdoba Azcárate.

I have been incredibly fortunate to forge so many strong friendships with my brilliant colleagues in the Department of Philosophy & Religion at the University of North Texas. I am especially grateful for ongoing collaborations with Irene Klaver as friend, mentor, and cothinker. Thank you for so closely reading every word of the book and proposal drafts, and for modeling steadfast support I can only aspire to reciprocate. Terra Rowe and Leah Kalmanson have read multiple chapter drafts and been there for me through the highs and lows; Miguel Gualdrón-Ramirez, Nathalia Hernández-Vidal, and Samantha Langsdale continue to inspire with their outstanding scholarship and integrity. At UNT more broadly, thank you to Lauren Fischer, Elinor Lichtenberg, and Brian O'Connor for your friendship and support. During pandemic times, work friendships and family merged in the form of the COVID pod. I cannot fathom enduring those years without the company of Terra, Mica, and Bea, or the many walks with Miguel, Nathalia, and

Antonia. I especially owe my thanks to Jim, without whose quiet generosity I may very well not still be employed. Emerging from pandemic survival mode was made vastly more manageable with the structured support of the Superpoderosas Writing Group, tackling writing challenges one pomodoro at a time: my thanks especially to Diana Pardo Pedraza, Soulit Chacko, Chinbo Chong, Manuel Cuellar, David Tenorio, and Leniqueca Welcome.

I have grown in unmeasurable ways sharing conference panels, round-tables, and symposia with outstanding scholars whose work has become deeply entangled with my own. In particular, I would like to thank Stefan Helmreich, whose exemplary research and generosity continue to resonate deeply; and my cothinkers with islands, Josh Reno and May Ee Wong, who have both read draft chapters and engaged in many generative conversations. In thinking islands with STS, I thank Seung Hee Cho, Josefina Arriagada Poblete, and Riley Taitingfong; for mediated seas and colonialism, Kate Sammler, Ben Mendelsohn, and Lisa Han; for oceanic humanities, Helen Rozwadowski and Laura Winkiel; and for the shapes of things: Lydia Gibson, Tone Walford, Donald McNeill, and Lourdes Vera. For astute questions and comments, many of whose prescient impact I would not come to realize until far later, I thank Marisa Brandt, Xan Chacko, Matei Candea, Joe Dumit, Jessica Jordan, Roshanak Kheshti, and Miri Powell.

At the MIT Press, I extend my thanks to editor Justin Kehoe, for his kind guidance and deft navigation of institutional processes, and to Suraiya Jetha for wrangling so many details. The manuscript has been immeasurably improved by the generous, constructive feedback of multiple anonymous reviewers. In addition, portions of the introduction and chapter 3 were developed in previous essays and with crucial insights from their respective reviewers: the 2017 article "Plastic Naturecultures: Multispecies Ethnography and the Dangers of Separating Living from Nonliving Bodies," *Body & Society* 23(3): 23–47 and "Plastivores and the Persistence of Synthetic Futures," in *Living in the Plastic Age,* Johanna Kramm and Caroline Völker (Eds), Campus Verlag.

My own circulations around the Pacific would not have been possible without funding from the National Science Foundation; UC Humanities California Studies Consortium; the University of California San Diego

Department of Communication, Science Studies Program, Japanese Studies, Institute for International, Comparative and Area Studies, Global California Studies, and Dean of Social Sciences; and the UNT Department of Philosophy and Religion.

Finally, and for always, an ocean of enduring gratitude to Stephen Mandiberg, who has read and formatted some of these sentences more times than I have crossed the Pacific. Thank you for your unwavering support through daily struggles mundane and spectacular; for sitting by my side through the longest nights until the wee hours grew larger again. And to Jade, for approaching every day with exuberant curiosity and kindness, and for giving the best rainbow bug hugs.

INTRODUCTION

Sail through the North Pacific Ocean and surely you cannot miss it: the Great Pacific Garbage Patch. It is massive and growing. An endless flow of plastic waste spun tight into an island by a spiraling oceanic gyre. Crowned the World's Largest Landfill, the garbage patch contains an estimated 1.8 trillion pieces of plastic weighing 3.5 million tons, spanning hundreds of miles, and extending up to one hundred meters deep.[1] It is a shameful monument of consumer excess so solid you can walk on it. You can stand upon its synthetic shores. Everywhere you look is plastic. Iconic water bottles, bags, and cups pile up amid lost fishing nets and nylon ropes, all tangled with countless scraps recognizable only as synthetics. Step with caution as you survey the island's shifting edges where a murky halo of microplastics and toxins leaches into the blue surrounds of the sea. Gaze down with the eyes of satellites and the immensity of its contours becomes visible from space: an entire eighth continent where there should be only ocean. Declare it the most shocking thing you have ever seen. The Great Pacific Garbage Patch has grown so large it is twice the size of Texas.

Except, no one can find it.

The trash island is not there.

And yet, you would have good reason to believe that it is. Nearly every claim above has been published by a major news outlet, if never all of them at once.[2] Captivated by visions of a plastic island, in July 2011, I set sail with a crew of researchers, journalists, and activists on an expedition through the garbage patch organized by California nonprofit Algalita Marine Research and Education. Our journey and my thinking since have been defined as much by what we did not see as by what we did encounter. Humbled by a tumultuous first week of rough seas, the calmer weather system that announced our arrival in the heart of the accumulation zone brought physical relief, but little plastic. I gazed out from the boat deck, right where the garbage patch was supposed to be, incredulously surveying an expanse of seemingly pristine blue ocean extending to the horizon in every direction. The crew was plagued with doubt. "Are we there?" I joked daily with my shipmates, who shrugged and laughed in response. It was only by towing sampling nets alongside the boat that we were able to see what was invisible from the deck above: a confetti sprinkle of fragmenting plastic bits mixed up with small sea creatures. The scientific process meticulously brought into being a garbage patch not as a solid trash island but as something far more fluid, contingent, and teeming with life.

We were in the garbage patch to do science and to see for ourselves. Yet, the storied island has refused to give way to measured observations. Near the end of the expedition, after three weeks at sea, some onboard attempted to explain what we had missed. One crew member proclaimed the island a "ridiculous notion," then immediately qualified that it was very unfortunate we did not see "the true heart of the accumulation zone," as its existence could supposedly not be denied. Another participant reasoned that if you added all the bits together there would be enough for an island. Back on land, I have continued to watch similar scenarios play out again and again, as scientists and educators present audiences with data, reports, images, and sample jars of tiny plastic shards suspended in seawater—accumulating evidence that an island does not exist—only to be bombarded with suggestions for towing "the island" to shore for recycling, or for converting it into artificial ice floes for polar bears. Despite the tireless work of those producing and circulating scientific knowledge about ocean plastic pollution for decades now, many people still struggle to reconcile tiny plastic fragments scattered across the sea with continental expectations.

The trash island is no mere misrepresentation. Accounts of science communication grounded in accuracy cannot explain it away. The island is so powerful that it continues to persist in the face of tireless attempts at clarification and a whole spate of alternative descriptions: plastic soup, plastic smog, garbage belt. So powerful that some journalists argue climate change needs an equivalent rallying representation. So powerful that when confronted with the lack of a trash island, some architects insist that one must be built. Even in its absence, trash island continues to shape popular understandings of and dominant solutions to ocean plastic pollution, and to environmental crises more generally.

This book is about what the trash island—that simultaneously does not exist and will not go away—can tell us about the persistence of plastic pollution and all its associated harms. It is a story about how knowledge and awareness of global ecological problems so often fail to instigate meaningful change; a story where plastic pollution becomes a new substrate for propagating the very relations that led to its emergence in the first place. It is a story where environmental crises become synthetic frontiers: place-making expansions of petrocapitalism that keep producing colonial landforms. The garbage-patch-as-trash-island is the consolidation of dominant ways of knowing and exploiting a Pacific Ocean constituted by elemental land/ water divides and the fluidities that defy them. I show how an ocean emptied of human presence becomes a space of nature for expansion, exploration, exploitation, and plastic contamination, which in turn becomes a synthetic frontier for further colonization and profit. Re-turning around and around the persistent plastic present of my journey through the garbage patch, this book embodies the recirculating relations of oceanic gyres, where a trash island very well could, but does not have to, come to be. With the fluidity of elemental substances in their many forms, there remains still the possibility of refusing to shore up the violent divides of synthetic frontiers, to imagine the kinds of relations with oceans and with each other that might be built instead.

SYNTHETIC FRONTIERS

The plastic trash island has eerie precedent in plastic industry aspirations that prefigure garbage patch science by many decades. Plastic was being positioned as a defining substance well before human agency was conceived

Figure 0.1
The map of "Synthetica: A New Continent of Plastic" as it appeared in *Fortune Magazine*, 1940. Note the islands of Rayon and Nylon growing off the coast.

as having the kind of geologic force that defines the current Anthropocene epoch, and before associated harms were understood as planetary environmental crises. American journalists had already heralded the arrival of the "Plastics Age" in the 1930s, with a proliferation of new types of synthetic polymers, even before 1950s postwar consumer disposability and excess.[3] In a 1940s *Fortune* magazine article, the state of the plastics industry takes on explicitly continental form: a map of "Synthetica: A New Continent of Plastics" (figure 0.1).[4] Colorfully delineated countries and geological features, each molded from their respective namesake material by Ortho Plastic Novelties Inc, dominate a two-page spread. Nylon. Melamine. Vinyl. Lakes and rivers of acetic acid course without irony toward a translucent sea of rippled blue glass, assumedly immune to acidification itself. This era transformed by synthetics has a decidedly modern capitalist geography where borders are mapped with molecular precision and state power and corporate rule collapse into branded cities: Bakelite, the capital of Phenolic. "Cellulose is a great state, something like Texas," boasts the legend, before introducing Rayon as "a plastic island" off its coast.

As Synthetica melds the frontiers of chemistry (new kinds of molecules synthesized in the laboratory) with the frontiers of colonialism (new territory for political control and settlement) and capitalism (new products for expanding profits), plastic becomes new land.[5] Shaped into an amalgam of the South American and African continents, Synthetica emerges as territory not only already claimed for, but whose very existence *as land* emerges with resource extraction and elemental manipulation. At the same time, the chemical-political boundaries defining the new continent are explicitly "unsteady," anticipating the shifting frontiers of chemistry's continued advances. With the island of Melamine marked as "new territory," Synthetica is only expanding. That such rapid chemical industry growth was a continuation of the westward march of progress was articulated by industry leaders themselves. As cultural historian of plastic Jeffrey Meikle astutely details, industry "pioneers" at the time explicitly "looked to horizons as inviting as those facing the explorers who opened up the American continent," with one industry executive likening the realm of plastics to a "wild and wooly West."[6] Frontier, in this sense, is more than synonymous with boundaries to be exceeded; it cojoins the cutting edges of western science and technology with colonial movements toward a Pacific Ocean wilderness.

Synthetica makes visible, even celebrates, the relations to place that have continued to shape ocean plastic pollution into terrain for further growth. Yet it is through the active denial of plastic terraforming that future trash islands become possible. As Meikle qualifies, Synthetica represents one polar extreme of competing plastic imaginaries at the time: plastic as an extension of the natural world, as opposed to plastic as a liberation from it. Visually aligned with continents of the global south, the map geographically grounds plastic as a contiguous expansion of existing forms of resource extraction. As the legend narrates, "the countries march right out of the natural world— that wild area of firs and rubber plantations, upper left—into the illimitable world of the molecule." Despite the contrasting textures of plastic and glass that materialize coastlines as an elemental divide—plastic as land, glass as the sea—the accompanying text attempts to coax these substances into a seamless trajectory of progress: Glass, asserts the legend, is "the oldest plastic known."

When it comes to dominant plastic imaginaries, however, Synthetica's antithesis appears to win out: plastic as radical liberation from the limitations of natural materials and the caprices of geography. The decades that follow are best characterized by a plastic culture of "otherworldly escape" and "proliferating transcendence."[7] Plastic objects seemingly appear out of nowhere and, upon disposal, are magically thrown "away." They bear no traces of their extractive origins and betray no hints of their geological afterlives. In the 1950s, the material so unnerved Roland Barthes with its alchemy of everything and nothingness, he described plastic as "the very idea of its infinite transformation . . . ubiquity made visible."[8] This material transcendence is exactly the kind of active disconnection of plastic from its relations with specific places, people, and power that culture and media studies scholar Heather Davis calls out as synthetic universality. Positing that plastic matter "embodies the Western desire to rid ourselves of our obligations, relations, and connections to the land," Davis joins contemporary scholars in drawing attention to the violences that make plastic's very ubiquity possible.[9] Plastic capitalism animated by oil endlessly turned object.[10] Plastic pollution enacting settler entitlement to Indigenous land for storing waste.[11] Plasticity a technology of slavery forcing Black bodies to push the limits of what it means to be human.[12] Where, as these scholars have traced, Western dreams of plasticity long precede polymers derived from fossil fuels, so too do the promises of synthetics.

I join those taking up the work of uncoupling synthetics from the violence of universality by insisting on keeping synthetics in relations to place. By synthetic I mean the constitutive melding of knowing and making that precedes and exceeds modern plastics.[13] While synthetic has become synonymous with artificial, unnatural, even fake, its oldest traceable roots are the ancient Greek words for "place" and "together." Synthetic shares these origins with synthesis, describing methods for skillfully making something new through the combination of component parts.[14] Returning emphasis to making and placing together keeps attention on specific practices that bring elements, forms, and even whole continents into being. These are processes where it very much matters on whose terms component parts are defined and whose new worlds they build. Despite the strong contemporary resonances with materials molecularly forged in the laboratory, the term "synthetic" first entered the English language to describe the processes of Western philosophy, not the products of modern chemistry. On the eve of the seventeenth century, synthetic philosophy described ways of thinking that proceeded from most simple principles to higher truths.[15] Synthesis in particular has become strongly associated with the Hegelian dialectic, particularly as taken up by Marx and Engels, as the resolution of tensions into ever more sophisticated stages of existence.[16] Associated with the inevitable evolution of the modern subject, such conceptualizations of development are violently enacted, as philosopher Brian Burkhart points out, with the conquering of "undeveloped" peoples and landscapes.[17] Hegel's ontology is "geographically linear," unfolding in space from Europe to America, a new land to be made fully formed by European expansion: the westward movement of the American frontier acting on "the passive and natural."[18] As with plasticity, synthesis gets caught up with severing existing relations to place to impose designs on others.

With nineteenth-century chemistry, synthesis had come to describe scientific methods for knowing nature by making rather than analyzing materials.[19] Synthetic chemists strove to combine elements of the periodic table variety to create new forms of matter, just as philosophers combined principles to produce higher truths. As most commonly recounted, the first fully synthetic chemical substance was not plastic, it was purple.[20] A dye that that British chemist William Henry Perkin formed by accident from coal tar—the waste from processing a fossil fuel to power industrial expansion. With quinine in high demand by the growing British empire, he had been trying to create a cheap treatment for malaria, but instead made mauve. It is no accident that, soon after, the synthetic chemical industry flourished in Germany: a country with plentiful coal but notably short on colonies. As political theorist Ester Leslie compellingly argues in her book *Synthetic Worlds,* dialectic philosophy laid the foundations for the dreams of plastic matter as chemical reactions too "bring opposites together in an exchange of properties to produce new things."[21] Moreover, by turning coal-based waste into matter of value, chemical synthesis constituted new territory for expansion in the absence of access to empty land: "Where space—or the right space, the right land—was lacking, science could step in to compensate."[22] Endless material substitutions beget synthetic empires of molecules.

Synthetic frontiers, then, are place-making expansions of petrocapitalism and Western science that perpetuate colonial land relations. They consolidate modern dreams of endless molecular manipulation and territorial power into the conquering of ever-more matter in landform.[23] Synthetic frontiers sometimes look like literal plastic geographic features—the continent of Synthetica, a garbage patch trash island—and at other times, like microplastic habitats colonized by bacteria now heralded as the missing link in a circular plastics economy. The trash island as synthetic frontier already figures in fiction, vividly illustrated with graphic novel series *Great Pacific* terraforming plastic trash with unmistakably American colonial energy (figure 0.2).[24] The cover image from "Part Three: Nation Building" in particular manifests an uncanny reincarnation of the 1940s plastic industry continent, a garbage patch island of colorfully delineated states claimed in the name of New Texas (note the lone star flag on land and the Pacific Islander figure on the sea, all staged in a Risk-like game of conquest).[25] Yet, synthetic is no mere synonym for fake any more than frontier is a metaphor for exceeding limits. Synthetic frontiers physically mark and maintain boundaries across shifting spaces, with lines drawn on maps, inscribed into environments, lines that enclose oceans and cut entangled bodies. These relations render land not only as resource and property, but, in the case of ocean plastic pollution, as elementally distinct from the sea. Potential crises become sites that ensure further growth: seascapes, bodies, and even pollution itself molded into landscapes of persistent progress. Synthetica maps responsibility for the universalizing logics that give the garbage patch trash island its power.

OCEAN PLASTIC POLLUTION

The story of contemporary concern for ocean plastic pollution as it is most commonly told, does not exactly involve mapping a continent of plastic fully formed, but it does begin with a discovery at sea. In 1997, Charles Moore, a boat captain who used his oil inheritance to found the nonprofit now called Algalita Marine Research and Education, was sailing home from a race in

Figure 0.2

Cover image from issue 9 of Joe Harris and Martín Morazzo's graphic novel series *Great Pacific* depicting a trash island continent as gameboard. Two white men, one holding a lone star flag, stand in opposition to an Indigenous Pacific Island woman and a cloaked figure.

Hawaiʻi when he took a little-traveled shortcut that landed him in the doldrums with rapidly dwindling fuel reserves. Slowed to a crawl, he began to notice an alarming amount of floating consumer goods and plastic fragments littering the sea surface.[26] While Moore did not, as he clarifies often and with great emphasis, find an island, a mountain, or a vortex of plastic, the experience did compel him to redirect Algalita to focus exclusively on ocean plastic pollution. As Moore soon learned, he had discovered what oceanographers had already modeled, predicted, and named up to a decade earlier: an area in the North Pacific where converging ocean currents gathered and recirculated floating synthetic debris.[27] Oceanographer Curtis Ebbesmeyer had dubbed this area the "garbage patch," but it is also commonly referred to as the trash vortex, plastic patch, or even the garbage gyre.

The waste that congregates in garbage patches is overwhelmingly plastic.[28] When I use the term *plastic*, I am invoking the cultural category as it is used to contain thousands of heterogeneous synthetic materials.[29] This common usage of plastic is almost always shorthand for synthetic plastic, meaning materials produced through modern chemistry from fossil fuels and their industrial by-products, as petrochemicals caught up in the promises of "better living" that are at the same time problems of war, environmental injustice, and climate change. Not only has a massive amount of plastic been made since the mid-twentieth century, it has been made durable. All plastics are polymers, which means they are super large molecules made of extremely long, repeating chains of chemical units. While not all plastics are petrochemical (i.e., celluloid, which is derived from plants), and not all polymers are formed through human intervention (i.e., silk, which is formed by insects), the bonds linking industrial plastic polymer chemical chains are too strong or too unfamiliar to be broken down by organisms aside from a few exceptional microbes and fungi,[30] hence, plastic's incredible feats of material endurance.

Most of this plastic is not dumped in the ocean intentionally through malicious human acts of littering. Plastic spills as raw materials from production facilities and trains, flies out of garbage cans on collection day, sheds from laundry, and washes into storm drains, down creeks, and off beaches. From a relational perspective, however, individual intentions do not excuse actions that cause harm without the consent of those whose bodies and environments become contaminated. Both passive voice descriptions (plastic was found) and the active transferring of all agency to plastic materials themselves (plastic causes pollution) have a tendency to lose track of relationships while shifting responsibility elsewhere. Most notably, the focus on consumer bottles, bags, and straws that have become icons of plastic pollution bolsters assertions that consumer choice is where things go wrong. In contrast, recognizable objects in the open ocean are far more likely to be industrial, whether lost fishing gear, aquaculture supplies, or the beadlike preproduction plastic pellets nicknamed "nurdles" that become waste without ever being used. Moreover, not all types of plastic float in seawater. With the cap off, even single-use plastic water bottles will sink to the sea floor. To constitute ocean plastic pollution as a garbage patch is to already limit the conversation to the subset of synthetic plastics that float, and with them, to focus on surfaces rather than depths.[31]

Once in the ocean, floating plastic can get caught up in giant circulating current systems called "gyres," formed by wind patterns and the very rotation of the earth. Five major oceanic gyre systems exist on the planet: the North Atlantic, South Atlantic, North Pacific, South Pacific, and Indian Ocean Gyres. Within these current systems are calm areas where anything that floats tends to gather. These are the garbage patches where plastic accumulates. A garbage patch is shaped by and located in a gyre current system, but at least in scientific terms, it is not the gyre itself. To put this all into practice, take the map of the North Pacific provided by the National Oceanic and Atmospheric Administration (NOAA) in figure 0.3. The arrows trace the movements of the major ocean currents, with those encircling Hawai'i constituting the North Pacific Subtropical Gyre. For simplicity, educators and journalists often locate the garbage patch in the middle of this system without being too specific, but NOAA's map demarcates two distinct areas of accumulation, complicating the image of a single garbage patch or island. The one to the east (cardinal, not cultural—thank you, Euro-Atlantic centrism) between Hawai'i and California is labeled "Area of the North Pacific Subtropical High or Eastern Pacific garbage patch"; it corresponds with the infamous Great Pacific Garbage Patch.

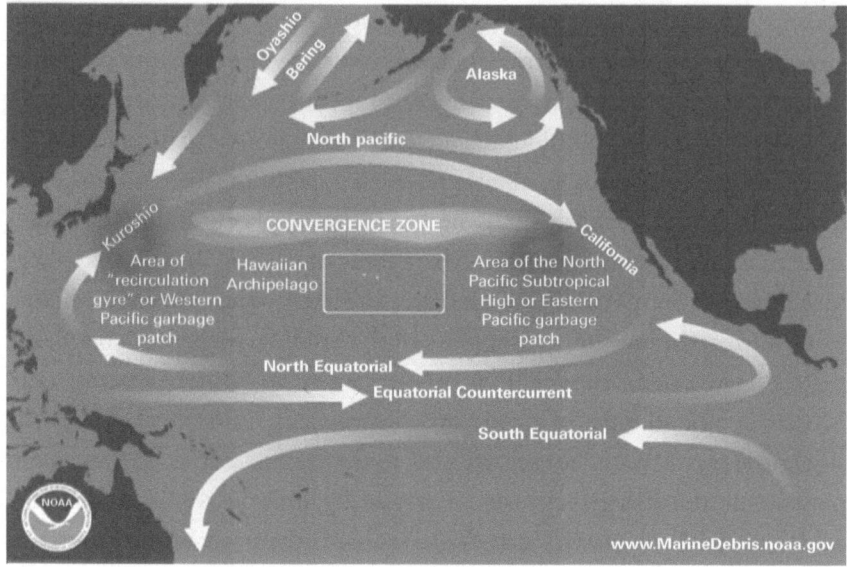

Figure 0.3

NOAA's map of the "garbage patches" in the Pacific Ocean. The area marked "Eastern Pacific garbage patch" is much more commonly known as the Great Pacific Garbage Patch.

As it travels with these gyre currents, plastic transforms and is transformed. With bonds susceptible to sunlight and mechanical fracturing rather than metabolic processes, plastics photodegrade (rather than biodegrade) into ever smaller pieces without disappearing completely into original chemical parts, a process made even slower in seawater—hence, the confetti fragments we saw at sea, more formally known as microplastics.[32] As they travel and break down, floating plastic becomes a multiple threat: through the entanglement of marine life caught by larger objects like lost fishing nets; through the ingestion of plastic as marine life treat items including lighters, bottle caps, and smaller fragments like food; and as a vector for toxins as plastics both leach chemicals into and concentrate toxins from the surrounding seawater.[33] Many of these substances, with ominous modern acronyms like DDT, PCBs, and BPA, are endocrine disruptors capable of tricking hormones and rejigging reproductive systems in even the smallest dose. For these reasons, and with emphasis on their potential economic consequences, the United Nations Environment Program (UNEP) named

"Plastic Debris in the Ocean" among major emerging global environmental challenges in 2011.

All kinds of simmering politics are caught up in such precise descriptions and careful definition of terms. This is especially evident in the tensions between the terms "marine debris" and "plastic pollution," particularly as they surfaced at the Fifth International Marine Debris Conference in Honolulu, an event hosted by NOAA and UNEP in March 2011. As recounted to me by Bill Francis, Algalita's then president, tensions at the conference coalesced around the "Honolulu Commitment," a two-page document meant to codify agreement among the many government, corporate, and nonprofit stakeholders in attendance. It generated just the opposite effect. To many participants' dismay, the draft document had a surprising omission: It made no mention of plastic, though the vast majority of panels and presentations were about exactly that. Even the official vocabulary of event organizers and sponsors exhibited a calculated avoidance of the words "plastic" and "pollution," especially in combination.[34] In the words of Plastic Pollution Coalition cofounder Daniella Russo, plastic was one "Who Must Not Be Named."[35] Instead, the document mentioned only "marine debris."

Not all marine debris is plastic; but, as Francis, Russo, and even UNEP have pointed out, most of it is. Plastic pollution, then, has become the term for those making an activist statement that ideally keeps plastic in its relations, assumes there is at least potential harm, and is followed by action. Marine debris, in contrast, has become the phrase of choice for those attempting to present a neutral stance, to signal a commitment to accurately describing what is or is not in the ocean in a way that is meant to be detached from particular courses of action. The conference, it so happens, listed as its two major sponsors none other than Coca-Cola, a company thoroughly invested in disposable plastic production, recycling, and greenwashing, and the American Chemistry Council (ACC), the trade industry association for plastic producers. The incommensurability of these approaches leaves its traces in the Honolulu Commitment. After many hours of negotiation at conference working sessions, the editors agreed to note the relationship between marine debris and plastic pollution. This concession was bartered in exchange for allowing the plastics industry to flaunt their plan for reducing the amount of plastic pollution (the result of which, a decade later, looks a lot like expanding plastic production). The final document makes a single reference to "plastic debris," followed by a much longer qualification "recognizing that other materials also constitute marine debris" (UNEP 2011). In very intentionally evoking ocean plastic pollution, I name a problem that demands actively working against practices that attempt to contain plastic by separating it from forms of responsibility.[36]

PLASTIC PERSISTENCE

The persistence of an ocean plastic trash island that refuses to dissolve in the wake of an ever-growing mass of scientific research is, at the same time, the story of propagating plastic extraction, production, and waste. The stubborn island image and relentless petrochemical expansion are connected by a whole constellation of relations constituting multiple sites and scales, connecting bodies and environments, pasts and futures, in circuits of differential harm. The solution to ocean plastic pollution is arguably very simple: Make less plastic. Understanding why this appears so impossibly out of reach requires an examination of what, exactly, "we" are trying to solve, and if "we" even want to fix it.[37] Simply banishing plastic materials does not attend to the relations that led to the very possibility of ever-expanding synthetic plastic production, relations that must be understood as part of the problem. In focusing on plastic's persistence as a question of what relations endure and how, I show how the trash island and dominant responses to it alike are caught up in the stabilization of the status quo of a planetary plastic crisis.

In the decade plus since I stepped aboard a sailboat journeying into the North Pacific, there has been an exponential increase in awareness of ocean plastic pollution. Innumerable organizations are now dedicated to ocean plastic research, education, and policy. This rising tide of concern has surged upstream, onto land, and into everyday lives, with promises of plastic-free oceans, rivers, lifestyles, and futures. Collective action arguably now constitutes a global anti-plastics movement, which some have declared a "worldwide revolt" or even a "war" against plastic. The global #BreakFreeFromPlastic movement alone boasts 2,831 member organizations.[38]

Recognition that plastic is "a planetary crisis" reverberates in the halls of the United Nations Environment Assembly, which voted to establish a global plastics treaty by 2024, with the understanding that there is ample scientific grounds for policy change.[39] Indeed, the UN global science report tracking research trends shows that the growth of scientific research on ocean plastic pollution far outpaced research on all other sustainable development-related topics between 2012 and 2019.[40]

Yet flows of plastic into global oceans are expected to increase, as petrochemical production continues seemingly unfazed.[41] Charting an accelerating trajectory of growth, global annual plastic production has increased from two million metric tons in 1950, to 380 million metric tons in 2015. Based on most recent trends, analysts project that in the year 2050, annual plastic production will reach 1,606 million metric tons, and plastic contributions to anthropogenic greenhouse gas emissions will have risen from 3.4 percent to a full 15 percent. Petrochemicals overtake vehicle fuel as the main driver of oil demand.[42] As a transition toward clean energy catalyzes demand for electric vehicles, the fossil fuel industry is seeking salvation with unprecedented investment in expanding petrochemical, and especially plastic, production. Researchers monitoring this investment by the twelve largest petrochemical companies (many themselves the world's largest oil companies like Exxon-Mobil and Shell or their subsidiaries) counted the announcement of eighty-eight new production and infrastructure projects between 2012 and 2019.[43] Alongside this growing capacity, the plastics industry continues mammoth efforts fighting anyone and everything that threatens the right to produce ever more waste in perpetuity. There is a willingness to do almost anything to assure the continuation of plastic's accumulations.[44] The petrochemical industry appears to be having the last laugh, reinvigorating the 1967 film *The Graduate*'s iconically hollow but seemingly unshakeable declaration that "there's a great future in plastics."

Trash island and plastic pollution are part of the same paradox of seemingly immovable endurance despite a flurry of activity aimed at addressing it. As a growing body of environmental and energy humanities research demonstrates, solutions deployed in the name of sustainability far too often become "green colonialism" and yet more extraction.[45] For anthropologist

Joseph Masco, this capacity to call attention to environmental dangers while simultaneously perpetuating their very conditions of possibility is inscribed in the logic of contemporary global crisis itself. Crisis, as Masco asserts, has become "a counterrevolutionary force," where appeals to urgency elicit a scramble to preserve the current way of life in the face of immediate threat, at the expense of considering—and actually addressing—the relations that led to the problem in the first place.[46] The result is a whole regime of stabilizing practices "emphasizing urgency and restoration over a review of first principles and historical ontologies."[47] This is an ocean plastic crisis that calls for containment, cleanup, and recycling without attending to continued production or to assumptions about the kinds of worlds that can or should be made plastic and at whose and what's expense.

While Masco writes in most detail of atomic and climate dangers, here too ocean plastic pollution is part of the same story, not only of stabilizing work performed in the name of global crises but of Pacific islands rendered waste, of carbon-emitting petrochemicals cracked into plastic resin, and the violences that join them. In the present moment, where these very connections are being articulated more and more forcefully—fossil fuel extraction as colonialism, petrochemical production as racial violence, plastic contributions to climate change—it is imperative that these powerful relations and calls to action are not siloed back into discrete issues. Narratives of separation lead all too easily to technological fixes that mitigate the threats to, but never the threats of, systemic harm. Plastic, as a substance wrought with dreams of modern control, seems particularly susceptible to this kind of compartmentalization, as it is so often reduced to a problem of mismanaged solid waste to be brought back in to circuits of production and value.[48] Already, in the process of drafting the UN global plastic agreement, a statement of concern addressing toxic petrochemical additives in plastics was removed at the insistence of delegates, including the United States, while the accompanying press release solidified plastic's place in "a new circular economy."[49]

Resisting such practices of containment, I join feminist science and technology studies (STS) scholars who insist on shifting focus from plastic as a material to plastic as relations. Where the interdisciplinary field of STS approaches knowledge and technology as inseparable from ongoing social and cultural processes, feminist work in particular holds tight to constellations of difference and power. As such, I trace the persistence not merely of a type of synthetic substance in the ocean but of plastic as part of a whole set of world building logics, narratives, and practices of domination. As defined by Heather Davis, plasticity is a defining modern orientation that assumes *all* matter exists to be endlessly manipulated for the benefit of (some) humans. At the same time, this namesake promise of malleability is founded on the violent imposition of modern visions of control on material substances, bodies, and land. For Davis, plastic's ubiquity depends on the active severing of accountability for these harms and, with it, obligations to specific peoples and places, as plastics are "designed and engineered to be universal, replicable, exchangeable, untraceable, and nonlocalizable."[50] Pieces of plastic could, but do not, list their chemical ingredients, place of manufacture, or fossil fuel source. The challenge, then, is grappling with questions of responsibility for materials that actively deny the relations in which they are embedded. Plastic's persistence is rife with paradoxes, of matter moldable into endless forms that are simultaneously resistant to change, of petrocapitalist growth where even the crisis of its own waste can be spun into new opportunities for profit, of a universalizing substance that clings to Pacific island form.

While very much caught up in the accumulation of wealth, the dominant relations that constitute ocean, plastic, and pollution are irreducible to capitalism. As Max Liboiron has powerfully established, the very existence of modern plastic is predicated on colonial access to Indigenous land.[51] The conditions of colonialism undergird the availability of matter to be manipulated, assuming entitlement to natural resources for extraction and to sites for containing the resulting waste. The Pacific Ocean in particular has figured in Western imaginaries as the ultimate sink according to ocean historians, whether almost too literally as a "capitalist basin[52] or in its immensity as a vast and therefore empty space of nature with the endless capacity to absorb pollution.[53] For Liboiron, modern pollution, and associated ways of knowing and managing it, are "essential parts" of the logics and practices of colonialism, no mere externalities or unintended consequences.[54] Moreover, the relations that produce spaces as nature (oceans empty of humans, land reduced to resource, material limits to be exceeded by synthetics), are far too often the unquestioned foundations for even well-meaning scientific and environmental work, which then fails to address perpetuating colonial relations. It is only by attending to the specific comings together of environmentalism with capitalism and colonialism in specific times and places that actions can have the potential to meaningfully resist reproducing intersecting forms of domination.[55]

The trash island that cannot be found turns out to be exceptionally good at making visible specific kinds of capitalist and colonial relations that constitute plastic's persistence in the North Pacific Ocean: the relations of synthetic frontiers. Awareness and actions are falling short (or working as intended), because of the forms of knowledge that are circulated and uncirculated, and the kinds of actions they render reasonable, even necessary. I do not mean that the majority of scientific research on ocean plastic has focused too much on describing the problem instead of solutions, as some science journalists have suggested.[56] I mean, as so many STS scholars have long contended, that science has never simply been describing. Claims about "what is" are inextricably caught up in delimiting what is possible or not. The very scientific studies I and so many others keep calling on to define the plastic crisis almost always work on unquestioned assumptions about managing synthetic waste, rather than not making it in the first place.[57]

Plastic's persistence is too often described as a stubborn physical quality alone: the prodigal endurance of a substance that, as commonly quipped, is made to last forever; of associated persistent organic pollutant forever chemicals; of claims that all the plastic ever made still exists somewhere on earth. The ubiquity of that somewhere is now performed by the obligatory listing of all the places plastic has now been found, namely in everything everywhere. Between material endurance and global presence, plastic has assumed force-of-nature status as a proposed marker for the Anthropocene, or rather, the Plasticene epoch.[58] These plastic "forevers" are part of universalizing human-oriented time scales and extractive geologics, not merely inherent physical properties. Endurance, as plastic excels at demonstrating, does not mean to last unchanging. Rather, plastic's persistence is also transformation, flows of chemicals, displacements of harms and responsibilities, and boundaries transgressed and forged anew. The trash island myth and garbage patch science alike emerge with practices of line-drawing that cut land from water, humans from the sea, nonliving from living bodies. These ontological and epistemological formations must be stabilized to ensure persistence, just as plastic seems to inspire other ways of thinking and living over and over again. By retracing persistent relations of power and obligation, ocean plastic pollution becomes re-embedded in specific sites of knowledge, environmental formation, and intervention. The relations—with matter, with land, and with the sea—that make plastic make sense also help make sense of why, of all the things plastic could become, it emerges as a persistent island of pollution in the Pacific Ocean.

ENTANGLED RESPONSIBILITIES

It is in the island-shaped shadow of a global plastic crisis that we headed out to sea. Faced with a seemingly endless expanse of empty ocean and dispersed plastic bits for day after day, when the very purpose of the expedition was to produce measured proof of plastic's pervasive presence, the appearance of a blue beverage crate adrift in the water just ahead instigated a flurry of action. The boat slowed, cameras rolled, and a team leaped into the water to investigate. To our surprise, the crate was home to a small school of tropical fish (figure 0.4). As first mate Dale swam back toward the boat pushing the crate, he joked that he was "driving the school bus." Though not physically blocked from leaving, the fish stuck with the container as if their lives depended on it.

As a coastal species that usually live near shore, the inhabitants have traveled a thousand miles into the open ocean, with the hard surface of the crate growing algae for the fish to eat and providing protection from predators. These particular fish could not exist in this particular place without the blue crate. The fish and the crate are what feminist STS scholars call *entangled*—their very existence is inextricably connected. They cannot be separated unchanged according to Western cultural assumptions of component parts of fish, plastic crate, and seawater. To speak of entanglements, then, is to invoke a relational ontology where the smallest unit of what exists in the world is the relationship rather than the individual. This is the difference between seeing fish-alive-with-plastic-in-seawater and a fish alone as a

Figure 0.4
Snorkeling with an algae-covered blue plastic crate of coastal reef fish in the middle of the Garbage Patch. Photograph courtesy of Algalita Marine Research and Education.

subject of a life in itself. Entanglements question the most elemental foundations of where a body ends and where the ocean or environment begins.

Conceptualizations of entanglement in feminist new materialism are about more than the inseparability of seemingly discrete material elements (though the term is often used in that way, particularly to denote a nature–culture hybrid ontology).[59] As specifically articulated by Karen Barad, entanglement is about "a fundamental inseparability of epistemological, ontological, and ethical considerations."[60] What the world is like, then, is inextricable from how we come to know it. As such, all knowledges are caught up with the prescriptive shoulds of ethics and power moves, of making truth claims that themselves make worlds. This stands in contrast to the map of Synthetica as a disembodied view of land from above, or to the language of marine debris claiming impartial accuracy. Instead, the knower is always situated, as Donna Haraway has long-insisted, always looking from somewhere.[61]

With the fish crate image, it matters that humans are immersed in the situation and immersed in seawater.[62] The entire scene is below the surface, light refracting toward the depths. Humans are part of the entanglement, with the explicit presence (ironically made possible by a plastic snorkel as lifeline to airy oxygen above and plastic mask prosthetic for human vision below), with the invisible presence of the photographer (with camera encased in a waterproof plastic cage), and with the community of "we" interpolated into being or (not) as readers of this book.

Moreover, feminist theories of entanglement help show how infrastructures and practices of observation have a tendency to produce that which they claim to merely measure. Ways of knowing ocean plastic enact and sever elemental and ethical relations; they change the world in their processes and set in motion actions that seem reasonable, even necessary. Though the purpose of the expedition was primarily researching changes in the distribution of plastic and the boat was not equipped for large-scale waste collection, touching the crate meant that leaving it in in the ocean felt like the moral equivalent to having tossed in there in the first place, as if somehow before that moment of direct contact we were not involved. Following conventions that as a human product, plastic did not belong in the sea, the team decided to remove the crate. Seeing the fish floundering on the boat deck before being unceremoniously pushed back into the sea without their shelter and food left me with the very unsettling sense that, at least in this particular instance, taking plastic back from the ocean was wrong. Encountering ways of living with plastic showed that taking plastic back as if there was some kind of giant "undo" button was truly impossible. With the crate, we could have plastic out of the ocean and dead fish, or live fish and plastic in the ocean together.

Western scientific ways of knowing tend to depend on categorical divisions between culture (plastic/humans) and nature (ocean/marine life), an understanding that readily supports conclusions that these substances should not mix, even more specifically, that plastic must be cleaned up from the sea. Yet in this case, taking away the plastic is to alter or even kill marine life. I do not mean to say that ocean plastic is good or to dismiss the suffering and violence it causes. The crate, left at sea, would continue to leach and concentrate toxins, eventually breaking into smaller pieces that could cause other forms of harm. Instead, I am calling on us to dwell with the discomfort of these fish severed from plastic as they undermine the assumption that cleanup is the right approach, or that aiming to keep human products separate from the environment is the right goal, or that nature-without-humans exists apart from practices that strive to make it so. As Barad elaborates, ways of knowing are conditions of (im)possibility, "they enact what matters and what is excluded from mattering."[63] Boundary-making practices, or agential cuts, simultaneously bring into being one possible reality while precluding others: the ocean as space of pure nature instead of considering better ways of living with the ocean and with each other. Thinking with entanglements is more than a fancy way of saying everything is connected; it is about taking responsibility for boundary construction where the stakes are sometimes deadly acts of exclusion or separation.[64] And I'm not just talking about the fish.

Considering entangled responsibilities with plastic begins with the acknowledgment that there is no outside of relationships from which to observe or evaluate environments. Or rather, that such outside positions are the product of agential cuts. The challenge is to start instead from the condition of being implicated (if never equally), whether conceptualized as compromise, complicity, or intimacy from which there is no going back.[65] As Davis poignantly writes of plastic, "This is not an escape from toxicity but rather a reckoning with its permeation."[66] This might start with the recognition of plastic not only as the crate but also as the snorkel, swimsuits, and camera casing our crew have brought with us and, most crucially, the synthetic molecules coursing through human, fish, and water bodies alike. But it cannot end there. Such approaches reject the possibility of some kind of moral high ground (with all its terrestrial baggage) or what Alexis Shotwell articulates as purity, the drive "to recover a time and state before or without pollution."[67] In the case of plastic, such purity politics might look like attempts to take plastic back from the sea after the fact, but also like the distancing language tricks of "marine debris," and like plastic-free living and zero-waste efforts to banish plastic objects from daily routines, as if buying a bamboo toothbrush can bring absolution. In arguing against purity, Shotwell makes a case for the impossibility of avoiding entanglements or untangling relationality, while attending to the not-so-innocent politics of purity itself as a goal. Here, Shotwell's argument echoes many others, particularly Bruno Latour's pronouncement that "we have never been modern," which points to hundreds of years of Western culture working constantly to maintain the pretense of nature/culture divides via practices of "purification" while spectacularly failing to do just that.[68] But what Shotwell does especially well is to connect individualized practices to broad systems of power by tying contemporary routines of cleaning, delineating, and forgetting to specific articulations of modernity and colonialism across science, culture and philosophy. As she explains: "To be against purity is to start from an understanding of our implication in this compromised world, to recognize the quite vast injustices informing our everyday lives, and from that understanding to act on our wish that it not be so."[69] In attending to the distributed agencies of new materialism, it is imperative to resist the dilution of responsibility.

To attend to what specific ways of knowing preclude, entanglement itself must be understood as part of entanglements. As Melody Jue and Rafico Ruiz contend, the term as it enters the humanities lexicon from physics, rests on decidedly terrestrial language of tangles as in knotted ropes.[70] These kinds of cultural associations lend themselves too easily toward understanding the fish crate as an entanglement *in* the ocean rather than an entanglement *with* seawater. The confluence of many elemental substances (including plastic in its chemical heterogeneity) in a comingled and especially aqueous state is arguably better described as saturation rather than entanglement. As defined by Jue and Ruiz, saturation "embodies the fluidity of relations that exceed attempts to contain, manage, and fix them to stable frameworks."[71] It is this tension that is precisely my point: how a specific terrestrial instantiation of matter-meaning brings ocean plastic pollution into being and then is surprised by its form. The language of entanglement does not itself end practices that cut, but neither does saturation (nor did hybrid, cyborg, coproduction, or dialectic). But thinking them together with a focus on exclusions and outsides can help recouple responsibility with the processes that construct synthetic frontiers, the place-making practices that render elements discrete, that cut bodies of/from water, that produce garbage patches and trash islands from cosaturations. The trash island becomes rogue object of myth and misrepresentation not to be debunked but to be immersed in the confluences of oceanic pasts and futures.

CIRCULATION AS METHOD AND SITE

Taking a relational approach means that instead of beginning with the assumption that plastic is an inherently "bad" substance, I look at where and how plastic emerges as part of contested material relations, as part of always ethical and political practices.[72] Or, in words of Jennifer Gabrys, how "plastics set in motion relations between things that become sites of responsibility and effect."[73] As such a site, however, the garbage patch is itself in motion, a circulation of circulations that seems to constantly evade attempts to anchor it with a grid of coordinates and facts and, much to marine scientists' chagrin, insists on making continued island appearances. As I tried to follow these moving relations, my research site became circulation itself—the comingled trajectories of all kinds of materials and practices: petrochemical-saturated currents of seawater, island formations, bodies, boats and sample jars, and human attempts to intercept, stabilize, and even disappear them.

My primary methodological entry point to understanding these relations was the much-recited STS imperative to "follow the actors,"[74] the humans and nonhumans through active practices that produce and circulate worlds of technoscience.[75] Stefan Helmreich's boundary-crossing ethnographic approach to ocean science in particular has been resoundingly influential in modeling how situated thinking with oceanic forms (whether waveforms, life forms, or seawater itself) can challenge the very terms of analysis, including the very "field" of fieldwork.[76] In my case, this began with an extreme introduction to participant observation on the boat crossing the Pacific, which became an STS right-of-passage laboratory study. I followed the samples collected at sea and helped separate, count, and otherwise transform them into data. These numbers, in turn, went on to become journal articles and then meta studies about the amount of plastic at sea, circulated as widely cited facts held up as evidence for the urgent need of action. I spent the bulk of the year after the expedition with Algalita in Long Beach, California, splitting my time between doing education work in their office and sample processing in their marine ecology laboratory. Algalita was at a major crossroads. In the years since Moore's garbage patch encounter, the organization had been working tirelessly to raise awareness. By the 2010s, the problem of ocean plastic pollution had reached the pages of the *Los Angeles* and *New York Times* and beyond, so why wasn't awareness equating to change? Though at the time, I did my best to gesture at systemic limits, pointing to the need for explicit political engagement (the organization at the time was staunchly antiadvocacy), this book is my much longer answer.

Ocean plastic does not simply transmogrify into facts and disappear. Following how it continues to move through bodies and environments, I interviewed scientists, activists, and policymakers all around the North Pacific. I joined beach cleanups in Hawai'i, built ocean trash sculptures at a community center in Oregon, and attended ocean policy forums in Japan. As I circled the North Pacific by plane, boat, train, on foot, and through texts, I began to notice the tendency for my investigation to loop around in unexpected ways. While interviewing a Japanese analytical chemist in Tokyo, I was shown San Diego State University and Algalita mugs, gifts delivered in person by two other participants in my study. I dove into the history of mathematical models of ocean currents, searching for the first scientific report of plastic in the Pacific, only to land at the Scripps Institution of Oceanography a few miles from home. I relied on analyses of historical, scientific, and media texts to "follow" plastic to times and places I could not visit directly. Flying to Hawai'i for the second time, gazing down on the sea far below, I realized I was making laps of the Pacific in the same direction as gyre currents. Only then did I understand what I had meant by following a site in motion.

Not only were many of my actors traveling, sometimes to very inconvenient places like the sea floor or inside bodies, they themselves were not always stable or given in advance. In following visible actors, the kind of science that gets published and cited, the people whose exploits are making news, the plastic problems as a garbage patch, I came to understand how dominant narratives were being made and upheld, mapping the lines of my own trajectory across the Pacific until I ended in a place of beginning to understand the stakes that come with questioning the very drawing of lines at all. In sum, this book is the culmination of two years of intensive participant observation and interviews, a decade more of media analysis, and exists as a point of temporary solidification in the ongoing process of figuring out what it all means.

CHAPTER TRAJECTORIES

This book spirals outward from the middle of the gyre, circulating again and again but never unchanged. Following the trajectories of plastic-saturated currents, its form resists the linear unfolding of Western progress and refuses to resolve once and for all the question of the trash island's presence or absence (figure 0.5). It begins, here in the introduction, on a journey into the middle of things: to the heart of the garbage patch; with the tumbling initiation of a terrestrially situated body into oceanic "field" work; with the realization of being already immersed in plastic-saturated entanglements. The interstitial pieces that follow re-turn to the garbage patch expedition: to

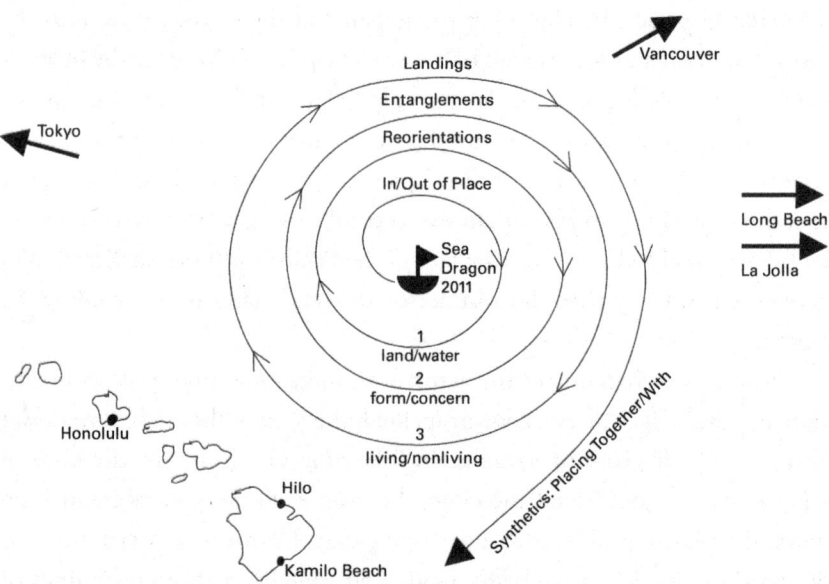

Figure 0.5
Map of book chapter trajectories in the North Pacific. Designed with Stephen Mandiberg.

the jarring embodied experiences of being simultaneously in and out of place, to the reorientations of an arrival that is not, where plastic and its lively entanglements remain to be seen. Each of the three main chapters between hones in on a distinct garbage patch formation as its sharpens the conceptualization of synthetic frontiers: The artificial coastline makes visible how synthetic frontiers work through the imposition of elemental lines as seemingly natural as the difference between land and water; the trash island embodies synthetic frontiers not as myths but as performative place-shaping processes; the plastisphere brings to life the evolving expansiveness of synthetic frontiers and the relations that persist. In the process, sharply delineated boundaries of separation and containment (land/water, form/concern, living/nonliving) dissolve into lively and deadly entanglements with plastic, opening up ways of radically reclaiming synthetics as collective practices of making place together.

Chapter 1, "Plastic Coastlines, Synthetic Frontiers," explores Pacific coastlines as synthetic frontiers: a Hawaiian beach awash with ocean plastic crumbling into a faded rainbow of synthetic sand; the nonprofit The Ocean Cleanup promises to restore pristine oceans by building "artificial coastlines where there are none." Through scientific mapping, a garbage patch arises with a hierarchical land/water divide enacted in decidedly Western terms. I develop a performative critique of terracentrism—the privileging of land over water—building from Dilip da Cunha's articulation of the line: Western colonial practices that bring land and water into being as discrete elements to be contained, owned, and extracted. Mobilizing feminist theorizations of difference and Kānaka Maoli (Native Hawaiian) relational responsibility makes visible other kinds of lines, other ways of land and water. Yet, following convergences of islands, waste, and Pacific frontiers reveals persistent relations between Western obsessions with discrete elements and bounded geographic forms. Legacies of colonial mapping, wastelanding, and technological experimentation continue as synthetic frontiers: The line expands with plastic to become new ground.

Chapter 2, "The Trash Island That Isn't There," follows the headlines of garbage patch science and communication. As elemental forms are entangled with concern, an ocean plastic pollution problem becomes inseparable from its island form. The trash island is a synthetic frontier that cannot simply be communicated out of existence. Expanding islands of trash figure as the stuff of headlines in the Science section of the *New York Times* and refuse to give way to supposedly more accurate descriptions. Attempts to myth-bust or to trace responsibility through linear origins stories are themselves quests that further entrench divisions between forms and concern at the expense of grappling with relations of power. As counternarratives circulate, knowledge of the trash island's absence does not make the problem of the persistent island—and especially the persistent plastic—go away. For the architects designing Recycled Island and the 8th Continent, the trash island's absence itself solidifies as a problem to be solved by *building* one. At the same time, plastic continues to escape scientific attempts to know and contain it on a global scale. Relations are disentangled to the point where a meta-study arrives at the surprising conclusion that "ninety-nine percent of ocean plastic has gone missing."[77] I show how the very existence of an ocean plastic crisis is inextricably connected to the generative potential of trash island's presences and absences and the enduring elemental lineages that make them powerful.

In the shadows of trash island lurks something far more tentacular: jellyfish with plastic embedded in their bodies; Bryozoa colonies coating plastic fragments; ocean insect *Halobates* not merely surviving but flourishing with plastic. Chapter 3, "Living in the Plastisphere," visits the newest synthetic frontier of the plastisphere: the realm of life forms that colonize and even metabolize plastics. Such persistent plastic entanglements across living/nonliving divides defy attempts to model, manage, and control them through elemental separations. Reading anti-plastic pollution campaigns against emerging plastivore science, I show how activism based on the same lines of separation that produce trash islands and garbage patches is especially vulnerable to technological fixes that maintain systems of harm. Just as many strategies for change continue to shore up divides by insisting plastic and bodies should not meet, they are being challenged anew with the emergence of plastivorous life forms. With the capacity to metabolize synthetics, these plastic-eaters are gnawing away at dominant understandings of plastic's material persistence. Yet, plastivores, by emerging as already coopted into circular plastics economies, at the same time threaten to save the status quo of plastic production and all its concomitant power relations.

Landings re-turns once more to the *Sea Dragon,* to the shores of empire, to the loops of transformation and persistence. In place of a conclusion, I propose radically reclaiming synthetics from both modern chemistry and modern philosophy to catalyze new kinds of elemental trajectories in the Pacific. Synthetics as *placing together/with* makes space for and with a multitude of movements that resist the linear progress of frontiers. Synthetics, as ways of making place together/with, does not presuppose discrete elements and refuses to resolve them into complete, bounded forms. Plastic-saturated oceans instigate new conversations joining the radical relationality of water, with anticolonial critiques of industrial chemicals, and with other ways of islanding. I envision possibilities for relations with oceans that no longer shore up violent natural/artificial or land/water divides.

IN/OUT OF PLACE

Hesitating, I contemplate the gap between sailboat and dock before stepping tentatively onto the seventy-two-foot *Sea Dragon* that is to be my home for the next three weeks. I move cautiously because I will share this space with twelve strangers, because this will be my first fieldwork experience, and because I have a tendency toward motion sickness, even on the swings at the park. I have never before set foot on a sailboat, and to cross this small slice of aquamarine harbor separating stable ground from rocking vessel is to admit that I am about to cross the vastly greater 2,706 miles of Pacific Ocean that separates Honolulu from Vancouver, all while supposedly conducting research and helping to steer a boat.

In the months leading up to the voyage, I poured over photographs and diagrams of the galley, bunks, and saloon posted on charter organization Pangaea Expedition's website, trying to imagine life at sea. With a blue-painted hull and a single mast standing almost one hundred feet tall, the boat in the harbor looks familiar, but my body struggles to navigate even the barely perceptible sway. I immediately stub my toes on one of the many rails, long metal bars raised a few inches off the deck that seem perfectly positioned for tripping. Heading below deck, I feel awkward on the ladder-step stairs, and wonder if the faint rumbling in my stomach is a sign of nerves, impending sickness, or both. I am assigned a bunk in the u-shaped communal sleeping area, the top tier of a triple. I survey the narrow stretch of blue canvas slung between two metal poles, the bib-like sides and dangling black webbing straps hinting that this space was not designed for gentle harbors. I soon learn that blue hulls are bad luck on boats. And so are women.

I meet the rest of the crew one by one, trying to match faces to photos from our expedition team bios. We have traveled from Australia, Canada, Korea, Taiwan, the United Kingdom, and the continental United States. We are teachers, researchers, artists, activists, and filmmakers, cautiously gauging experience, forging new friendships, and masking our fears. I am far from alone in my complete lack of knowledge about sailing. That I have company in my inexperience provides only the slightest comfort before congealing into calculated anxiety: We are about to embark on a nearly three-thousand-mile open-ocean crossing and the chance the person sitting next to me knows something about sailing hovers around one in four. The experiences we do bring are diverse yet united by shared concern for ocean plastic pollution and the desire to see the Great Pacific Garbage Patch for ourselves. We all have related projects on our minds, but it soon becomes clear that our first responsibility is simply staying on the boat.

Our noon departure from Honolulu is a highly choreographed media event, complete with a helicopter team filming from above. Plastic grocery bags accidentally brought on board with last-minute supplies are frantically spirited out of sight of the cameras as we pose in life vests and fragrant purple leis, smiles brimming with nervous energy. We motor out of the harbor, passing surfers at the break, waving to tourists on day cruises, and watching the high-rises of downtown Honolulu shrink behind us. Then it is time to raise the main sail, a somewhat hectic scramble as many of us do not know the ropes—the difference between a halyard and a sheet and what they control. As the wind catches the sail, the deck adjusts to a startlingly steep angle that is soon to become the new normal. Clinging to the deck, I suddenly understand the need for the toe-stubbing rails and tether lines that run the length of the boat. Media satisfied, we proceed along Oʻahu's protected leeward shores. I admire the mountainous coastline with new seriousness: This will be the last land we see for weeks. We take photos posing against the silhouette of the island as dolphins flip and dive at arm's reach, riding the wake of the bow. A rainbow appears in the mist of a distant squall.

And then we round the point.

As land quickly disappears into the distance, the horizon twists into staggered peaks and valleys fractured by white spray. Any lingering pretense of a leisurely afternoon dissolves into choppy ten-foot swells and near-gale thirty-knot winds of the open ocean so wrongly named Pacific. Without a stable point of reference, our bodies tumble inside out, succumbing to unfamiliar rhythms. It is rough enough that we are not allowed to vomit over the edge of the boat directly. Instead, we line up, crouching four in a row, heaving on the outer deck where the skipper stands unperturbed with the hose at the ready and the practiced reminder that if all we do between here and Vancouver is drink water, we will survive. Lying in our bunks in private misery, we are headed to the plastic accumulation zone, but we are for the moment completely incapable of observing anything beyond the plastic baggies clutched in our hands in case of further sickness. I close my eyes, as if gathering strength to endure the final thirty seconds of a spinning fairground ride. Except there are twenty more days to go.

There is a three-day gap in my field notes. What I remember of that time, like life on the boat, does not conform to terrestrial rhythms of night and day: a mess of dark and light dream states punctuated by moments of relative clarity. Fighting wave after wave of nausea; fumbling for gear in dark closets; begging for a water bottle refill; dreading trips to the toilet located in the treacherously bouncing bow. Though I try to pry my sea-tossed body from the safety of my bunk at the appointed time for watch that first night, my legs give way to vertigo: I am unable to stand. With shame and relief, I crawl back into bed where I remain for a full twenty-four hours. The first guest crew email to leave the boat via shared satellite account bears the telling subject heading: "Holy Shit Fuck Mountain."

At 2:00 a.m. on the second night, I am still feverish and disoriented, but with help from the team going off duty, I manage to get safety gear on my body and my body on deck. Ascending from the stuffy hold, the cool air is an instant relief; the Milky Way spills across the dark sky. The first mate decides it is a perfect time—the middle of the night, seas and stomachs still roiling—to teach a complete novice to steer. I'm given the current compass declination and instructed to aim for a star straight ahead. With a few tips for not overcorrecting against the waves, and warnings about something called "luffing," which I deduce has something to do with the sail making flappy sounds, the introductory lesson is over. I am left to direct our progress toward the garbage patch. Tethered to the boat by the five-foot length of red webbing attached to my life jacket harness, I am jostled by swells I cannot judge in the dark, which periodically send splashes of seawater over the bow and directly into my face. The wheel comes up almost to my chin, and feedback from the boat is occasionally strong enough for my feet to achieve liftoff. "Stay on the boat. Keep wind in the sails. Aim for the star," I whisper to myself as we fly along, bright streaks of meteors above, bioluminescence below.

1 PLASTIC COASTLINES, SYNTHETIC FRONTIERS

Kamilo Beach on the island of Hawai'i is the stuff of ocean pollution legend. As Pacific currents converge with the island's windward shore, ocean plastic piles skyward and crumbles into a faded rainbow of synthetic sand. The garbage patch accumulation is made visible in such spectacular form that in 2020, *The Guardian* dubbed the beach "one of the world's dirtiest places."[1] In Hawaiian, Kamilo means twisting currents, and these shores have long been known as a place where giant logs prized for canoe building would wash ashore.[2] Over the past few decades, locals have taken to calling it plastic beach. Organic materials are still brought here by the sea—fibrous coconut husks and tattered palm fronds, chunks of rough tree bark separated from smoothed branches, and a variety of seed pods—but I am here specifically to see plastic.

The synthetics reconstituting Kamilo's shores are at once familiar and strange. This is not litter in the expected form of crumpled wrappers left behind by careless beachgoers.[3] Relative to so many other Hawaiian beaches, Kamilo has few visitors at all. Accessible only by boat or by a long, bumpy four-wheel-drive road that crosses jagged lava fields, Kamilo tends to attract those searching for solitude . . . or for plastic. My guides for multiple visits in 2012 and 2013 are local beachcombers Noni Sanford and her husband, Ron, who in their incredible generosity have picked me up from Hilo, welcomed me to stay at their home in Volcano, woken before dawn, navigated the rough roads, and celebrated our arrival at the beach by sharing home-baked banana bread. In the early morning light, the beach is a stunning study in contrasts: bright greenery against distant rain clouds, colorful plastic against black lava exposed by the low tide. Walking the beach is a plastic sublime treasure hunt of nostalgia and horror (figure 1.1). For hours, we wander, eyes down, along the shore, digging through piles of plastic, synthetic confetti over a foot deep where it pools in the rocks. Much of the plastic that congregates here is weathered and fragmented, recognizable only in elemental terms—as plastic—rather than in its familiar object forms.[4] Because of this, we are drawn to the rarer recognizable items, especially those with traces of text that hint at origins: lighters imprinted with the names of businesses from around the Pacific, a role of green tape that reads "Tokyo gas," even a glass bottle labeled "U.S.S.R." Materials on the beach not only arrive from afar but also from a long time ago. They may have been circulating the North Pacific Gyre for decades before their paths again converged with dry land. Noni's own salvaged "toy collection" contains what, drawing on the fading memories of my own 1980s childhood, I recognize as a Battle Cat saddle from the 1982 run of Masters of the Universe toys; an enduring plastic form that is a fitting relic from a series set on the Earthlike planet of Eternia. Today, her archive gains a weathered red plastic figure that can only be described as an "Indian" in its stereotypical form and enduring displacement. I later tease a scraggly green swatch from the tangled tideline. My eyes widen with delayed recognition: it's AstroTurf. Synthetic grass. I am weeding plastic grass from plastic beach.

With such visibly plastic shores, Kamilo has become a poster beach for the problems of ocean plastic pollution. In addition to things that float at sea, it gathers beachcombers, film crews, researchers, and environmentalists, who all have sometimes disparate approaches to plastic coastlines. In my visits to Kamilo in 2012 and 2013, I walked the shore with Noni and Ron, but also meticulously sifted research quadrants with marine ecologists and cleaned up as much plastic as possible with an environmental restoration group. While I often saw the same people volunteering across these events (Noni regularly times her early morning visits to allow treasure-hunting before joining scheduled cleanups later the same day), their respective activities exemplified tensions about what, exactly, to do with the plastic. In particular, the environmentalist's urgency for cleanup to reduce immediate harm to wildlife was

Figure 1.1
Kamilo Beach at dawn. Photograph by the author. A fish-chewed plastic bottle is visible in the foreground, with bottles, lids, a dustpan, and many fragments mixed up with wood and other organic materials.

in palpable tension with university researchers who, while also concerned for marine life, wanted plastic materials left untouched for careful measurement before determining next steps. One cleanup crew was especially dismayed to find the beach covered in plastic bottles that oceanographers had recently intentionally released at sea for research purposes (the researchers in question later assured me they "got them all back"). Despite their differences, those who care most directly for Kamilo's shores are well aware that cleanup alone cannot stem the flows of plastic that seem as unstoppable as the next tide. At a Hawai'i Wildlife Fund event I joined in 2013, the organizer lamented that so many people who see images of the beach want to come to Kamilo to help, rather than considering how to take care of plastic relations in the places where they live. This invitation to contemplate specific relations to place continues to resonate as I reckon with my own complicity in colonialism and what, exactly, I have been doing studying plastic pollution in Hawai'i.

A decade after my first visits to Kamilo in 2012, beaches have not become less plastic. Instead, efforts to clean the ocean of synthetics are themselves begetting new plastic coastlines. In the case of The Ocean Cleanup, the award-winning nonprofit environmental organization that claims to be engineering "the largest cleanup in history," this involves developing technologies for extracting plastic from the Great Pacific Garbage Patch.[5] As many others have pointed out, The Ocean Cleanup is yet another massive techno fix that perpetuates plastic production by distracting from the political engagement needed for systemic change.[6] What is more specifically generative for conceptualizing synthetic frontiers is how The Ocean Cleanup prominently describes their massive project in terms of "creating artificial coastlines, where there are none."[7] These artificial coastlines are massive nets, suspended from long floating booms, themselves made of high-density polyethylene, a common type of plastic.[8] An early design envisioned these arrays firmly anchored to the sea floor to harness the natural energy of currents but was deemed impractical. Instead, the 2023 test involved a 2.2 kilometer (1.4 mile) long boom system towed between two boats, their fossil fuel emissions countered with carbon offsets (figure 1.2).[9] The Ocean Cleanup positions not only plastic coastlines as the problem but themselves as the *solution*: a

Figure 1.2
The Ocean Cleanup's System 03, an artificial coastline cleaning up plastic in the Great Pacific Garbage Patch. Photograph from The Ocean Cleanup.

critical component of their quest to "rid the ocean of plastic" that was backed by a fifty-three-million-dollar budget for 2022 alone.[10]

Positioned in the open ocean, these "floating barriers" are meant to capture plastic for removal by mimicking the way plastic washes up on beaches, visibly accumulating as it does on Kamilo. The design is rooted in The Ocean Cleanup founder and CEO Boyan Slat's own unexpected encounters with plastic coastlines. Slat was only eighteen years old when he gave the viral 2012 TEDx Talk, "How the Oceans Can Clean Themselves,"[11] where he recounts the organization's origin story: a diving trip to the "pristine" Azores islands, where he was shocked to see synthetics accumulating on the shore. In the TEDx Talk, Slat presents a glass cylinder of synthetic sand that could very well be from Kamilo beach as evidence of an environmental problem, before rapidly moving from specific plastic encounters to solving a global ocean crisis. The solution? A design for massive ocean cleanup arrays, with which he boasts the "Great Pacific Garbage Patch can clean itself in only five years." Slat's associated descriptions of the ocean as a watery "enemy" and of cleanup as "getting all the plastic back to land" morph into producing new landforms where plastic coastlines become something to emulate to restore a pristine sea, which leaves me asking: In what world is making more artificial coastlines the *solution* to artificial coastlines?

This chapter explores the synthetic frontiers of plastic coastlines as they delineate both the problems of, and most-hyped solutions for, ocean plastic pollution. Coastlines make especially visible how place-making expansions of petrocapitalism work through the imposition of Western elemental boundaries. Beginning with the terrestrial tropes of my own adventure-science experience in and out of place at sea, I explore how the garbage patch is caught up with what Dilip da Cunha call "the line": a technique of Western colonial power that brings discrete elements into regimes of domination and control. By bringing hydrohumanities critiques of terracentric land-water divides together with Pacific and island studies, I reconfigure the line of separation with an insistence on lines of radical connection and responsibility.[12] Situating the garbage patch as a continuation of the cutting edges of colonial expansion, I argue that the territorial power animating synthetic frontiers is enacted not merely through environmental but also through distinctly Western elemental forms of control.

ELEMENTAL WATERLINES

My conceptualizing of synthetic frontiers began on Kewalo Basin Harbor's cemented shores in Honolulu, though I was so preoccupied with the immediacy of my own aqueous fears that I did not realize it at the time. In the days preceding the departure of what is still the only sailing trip I have ever been on, I gazed suspiciously out to sea from the beaches of Oʻahu, full of nerves. When the time came to climb aboard the *Sea Dragon*, I hesitated. Toes perched at the edge of the solid concrete dock, the *Sea Dragon* was at once a single step and an impossible distance away. For a few seconds that felt like forever, I stared at the slice of turquoise water that separated me from the boat. The significance of this line to be crossed is so seared in my memory that, when I sought to use the image in a 2023 conference presentation, I was surprised to find that I did not actually have a photo of it (figure 1.3). I experienced the separation of land and sea as an elemental break so solid I

Figure 1.3
Contemplating "the line" between land and sea, aquamarine harbor and weathered gray concrete: a reenactment in Honolulu, 2023. Photograph by the author.

could trip on it. I had agonized excessively over what to pack for the voyage but was far less aware of the Western elemental baggage that I had brought aboard.

In his book *The Invention of Rivers,* landscape architect Dilip da Cunha draws attention to the long history of "the line": performative practices that bring land and water into being as distinct substances.[13] For da Cunha, land and water do not preexist as discrete elements: "This line does not simply separate water from land; it creates water and land on either side of it as entities that can be commodified and as such coveted, made scarce, and violated."[14] Just as Karen Barad's agential cuts constitute subjects and objects through exclusions,[15] da Cunha's line *creates* water from "fields of wetness" through separation and containment. It is with the line drawn on maps, embanked in soil, buttressed with concrete, that distinct elements are carved from fluctuating saturations and bounded in ways that can be owned, protected, or polluted. I performed the line with my hesitation, with my descriptions of water that slices, and with the language of an ocean that separates rather than one that connects. It is not the question of whether Oʻahu ends at the high or low tide line, but the assumption that the island ends at all: So many of us do not see the continuities of land as the sea floor, or of seawater as it rains into the soil from the sky.

Imposed on Hawaiʻi and so many other places, the line of Western land/ water is a technique of colonial power. It is also a pre-Cartesian dualism that joins empire and the empirical. Da Cunha traces the emergence of the line to a particular classical Greek worldview and ascribes its colonial expansion to Alexander the Great, whose conquering campaign, famous for its militaristic might, also counted among its ranks a brigade of scholars tasked with describing geographic features in meticulous detail. Their landscape-defining practices embody a particular form of vision that da Cunha refers to as "Alexander's eye," a god-trick of seeing land not only as a map from above but simultaneously as something drained by rivers that lead downhill to the sea.[16] It is from this perspective of the line—as Western elemental boundaries constituted and imposed—that da Cunha asserts rivers are inventions, not naturally occurring features to be discovered, but "acts of design."[17] Writing of the Ganges, da Cunha argues that the imposition of the line provides the foundation for the British colonization of India that "survived and thrived on keeping water contained."[18] The Ganges as a river, da Cunha declares, is "colonized rain." At the same time, the line that marks the river becomes a condition of possibility for understanding monsoon saturations as disasters. Without the line to be transgressed, without attempts to contain water in isolation, there can be no flood, da Cunha argues. Elemental boundary-drawing practices imposed become techniques of colonial power that bring "natural" forms of land, water, rivers, coastlines, islands, *and* associated crises into being.

Lines themselves are not inherently colonizing.[19] As feminist scholars— and especially Black feminist scholars—have long argued, the problem is not difference, but difference as the basis for separation that enables domination.[20] The trouble with *the* line is the way its cuts enact geographies of exclusion. Its binary separations bring into being the very ground for terracentrism: the long-standing Western privileging of land over water.[21] As an elemental form of domination, I understand terracentrism as enacted by performative practices that bring Western land-bounded perspective into being, while simultaneously positioning them as natural and universal.[22] Terracentrism pervades my notes from the *Sea Dragon*, rife with unmarked descriptions of water that breaks into "peaks and valleys," files and folders unironically labeled "field notes." Terracentrism is the planet Western conquerors continue to call Earth, when the ocean they bring into being and call Pacific has more surface area than all the dry land combined. It is modern forms of control that position water in opposition to much more valuable land, something da Cunha prefigures in describing Alexander the Great's forces so fearing the ocean that they changed course to avoid having to travel near the sea.[23] It is Boyan Slat of The Ocean Cleanup describing the ocean as an "enemy"—one that "can be used to our advantage" in returning plastic to (dry) land.[24] The line, as a colonizing logic that divides rather than connects, works to foreclose responsibility as it encloses elements in ways necessary for capitalist extraction.

Traditional Hawaiian land divisions have boundaries. Kamilo beach is located in the ahupua'a (community) of Wai'ohinu, in the moku (region) of Ka'u, on the mokupuni (island) of Hawai'i.[25] The main division, ahupua'a, takes its name from the words for cairn and pig, for the common boundary markers where offerings were placed. Hawaiian cartographies, however, map connectedness; they are not based on the line of Western land/water and all it contains. The very concept of land does not carry the same meaning. Though the word 'āina is often translated simply as "land," many contemporary Kānaka Maoli (Native Hawaiian) scholars are reclaiming its literal Hawaiian meaning as "that which feeds" or "lands and waters who feed."[26] When a Kamilo beach cleanup coordinator urged a school group to mālama 'āina—to care for their place—even if the plastic mess was not their own, she meant the mountain, shore, and ocean alike. 'Āina, "including land, sea, and society," by definition envelopes humans in relations of reciprocity with all kinds of kin, including plants, animals, and features of the land.[27] The Nation of Hawai'i's governance system was—and many are working for it to be again—configured from value and wealth rooted in water. Ahupua'a, as units of community life, are most iconically represented as wedge-shaped "pie slices" radiating from mountaintop toward the shore. Envisioned as bounded by ridges, there is a tendency to map them directly onto Western conceptions of watersheds: as land drained by the line of a single stream. Reproduced as a poster taught in schools across the islands, such standardized ahupua'a are recognizable by a group of Kānaka Maoli scholars, Noa Lincoln, Mehana Vaughan, and Natalie Kurashima, but do not resonate equally with their own lived experiences. Simplified into a singular Hawaiian resource management system, they argue that the poster version of ahupua'a homogenizes their shared diversity of place-specific relations. Fresh water is a defining element, but in so much more than Western watershed form; it includes "clouds, mist, groundwater, springs, dew, and rain . . . based on an understanding of the varied regional sources of water and the complex processes through which these sources nourished certain 'āina."[28] In contrast to the steep valley folds of rainier islands like Kaui'i, the leeward, dry side of the much younger island of Hawai'i is carved by springs and lava flows instead of perennial rivers. Here the boundaries of ahupua'a trace a diversity of shapes following clouds and rainfall, or even hop discontinuously, rather than being streamlined. At the same time, 'āina brings continuity with the sea. The diversity of ecozones gathered by ahupua'a include nearshore kai (sea), extending to outer reefs, or where the shores are steeper, for a distance offshore. And with it, embodied literacies of 'āina extend to ocean-based knowledges that Karin Amimoto Ingersoll calls "seascape epistemologies."[29]

Lines that bring worlds into being do not have to be grounded in domination. In her mapping of refusals of elemental foundations of Western power, Candace Fujikane beautifully makes visible Kānaka Maoli cartographies that map shared abundance through water, rather than scarcity through dry land. Hawaiian divisions, she emphasizes attending to smaller mo'o'āina land units, "are defined by their relationality with that which lies on their edges, borders that are not boundaries of separation but seams of relationality."[30] In place of oceans that separate, continuities of water above and below ground connect land across great distances. Moreover, "seams of relationality" trace relations of responsibility rather than severing them through enclosures that enable extraction. Kānaka Maoli, have a kuleana—a concept that at once encompasses obligation and privilege—to care for land. In the words of Mehana Vaughan recounting a conversation with her elder, with kuleana, identity is tied to "that which is yours to take care of." You become known by "how you use that to create abundance for others."[31] These lines are explicitly genealogical: Kānaka Maoli are descendants of the land with "familial relationality among humans, lands, and elemental forms, plant and animal 'ohana (family)."[32] Elemental forms are not inert substances to be controlled but teachers of how to listen, live, and care in relations of reciprocity.

In contrast, water's modern elemental form is deliberately unmoored from specific peoples and places, so that, like modern plastic, it can be subjected to new kinds of physical manipulation and containment. For geographer Jamie Linton, it is with abstraction from place that a multiplicity of situated, relational waters are actively reduced to a singular modern water, nature reduced to mathematical formula of hydro-logics.[33] For the nearly two thousand years that Aristotle's understanding of matter dominated in Europe, water was seen as an elementary, but also unstable and heterogeneous, substance.[34] When Antoine Lavoisier cracked water into constituent hydrogen and oxygen during the nascence of modern chemistry in the late eighteenth century, water as H_2O becomes not only a chemical compound but a substance understood to hold onto its essence even as it transforms in quality.[35] As such, modern water is made of specific cultural and political relations of nature/culture divides that simultaneously deny their role in bringing the world into being: "Modern water is *deliberately* non-social and non-historical in a way that the waters of other times and places are not," writes Linton. As water becomes modern, the imposed line becomes the stuff of progress, and Western elemental divides are cemented as foundations for further privileging particular kinds of geological forms.[36]

Water's situated and shifting status brings to question what exactly "elemental" means as humanities scholars turn to such approaches to subvert ever-powerful modern nature/culture divides. As outlined by Nicole Starosielski, elements understood broadly as "constituent parts" can be wielded to invoke substances or principles, building blocks or invariants (Aristotle's water vs. modern H_2O, for example).[37] Elemental philosophy is always political and ecological work as elemental lines imposed entangle powerful ontologies with territorial boundaries. On whose terms entangled worlds, watery embodiments, and co-saturations are cut into discrete substances and subjects must be made explicit.[38] Earth and water of the four Classical Greek elements do not map directly onto the five Classical Chinese elements; neither is dry land the same as that which feeds us.[39] If a river drawn with the line is colonized rain, then so too are islands drawn with the line colonized seascapes, plastic garbage patch islands included. The stakes are no mere intellectual exercise or cultural curiosity.[40] Holding the line of Western land/water props up infrastructures of containment as appropriate response in place of other kinds of relations with land and water, and other kinds of lands and waters. It matters whether and how distinct landforms and waterbodies are carved from fluctuating saturations, on whose elemental terms, and with what kinds of lines.

Plastic coastlines help make visible the territorial relationships between boundaries separating substances and those defining geographic forms: the elemental cuts of synthetic frontiers. The line is a specific kind of border that brings land/water into being as the foundation for expansions that are not only territorial but terraforming.[41] The line marks relations to place that delimit responsibility rather than enfold reciprocity. Practices of the line, with all its colonial elemental assumptions, are at work in both the scientific modeling of the Great Pacific Garbage Patch and in The Ocean Cleanup's project for addressing plastic where the solution is to "get plastic back to land." The kinds of elemental transformations that conjure colonial coastlines also consolidate plastic land from synthetic-saturated seawater.

CARVING GARBAGE PATCHES FROM SATURATIONS

Plastics are caught up with and concentrated by ocean currents, but they are brought into being as discrete, named, and even bounded garbage patch entities with the line of modern scientific processes.[42] The existence of a garbage patch in the North Pacific Ocean was predicted multiple times before it was "discovered" by Captain Charles Moore. In the 1980s, a team of oceanographers working for NOAA and the University of Alaska formally linked synthetic debris found in ocean surface samples with gyre currents, solving the mystery of how plastics were traveling so far from land and busy navigation routes: Plastic was being carried by currents, not dumped there directly by careless humans. In addition to noting dense concentrations of plastic particles in the Sea of Japan, the researchers inferred somewhat awkwardly that "the generally convergent nature of water in the North Pacific Central Gyre should result in high densities there also."[43] Such models linking plastics and currents are part of expansive modern projects for rendering the ocean manageable to control marine resources, especially fish. The North Pacific Gyre itself was named only once its currents were made calculable. Though the circular "motion of the ocean" was attributed to atmospheric patterns like the trade winds in the seventeenth century, and gyre patterns were described in the nineteenth century, terms like Subtropical Gyre first appeared as recently as the 1950s, after physical oceanography was firmly established.[44] As mathematical theories of fluid motion came to replace direct observations of water movement, they also formed the basis for the modern management of ocean ecosystems.[45]

Ocean current models themselves are developed, in part, by scientists throwing things in the ocean on purpose—first bottles and then satellite trackable drifters, but also by making use of more serendipitous items. Intrepid oceanographers (and also beachcombers), Curtis Ebbesmeyer and Jim Ingraham have been honing one such model first designed for predicting fish stock by tracing shipping container spills of plastic consumer goods back to their points of loss. The pair is best known for tracking the fates of nearly 30,000 world-traveling Floatee plastic bath toys but had previously honed their Ocean Surface Current Simulation (OSCURS) with a whole armada of beached Nike shoes.[46] Though shipping losses are usually secretive business, Nike's transportation department was surprisingly forthcoming, confirming that five shipping containers of shoes—a total of 78,932 sneakers—fell off cargo vessel *Hansa Carrier* in a May 27, 1990, storm mid-journey between Korea and Los Angeles (the contents of another sixteen containers lost at the same time remain a mystery). Since each Nike was stamped with a unique identification number, it was possible to trace individual shoes back to a specific point of entry into the Pacific. Floating sole-up with shoe bodies acting as rudders, lefts and rights had a tendency to take separate paths (a divergence that Ebbesmeyer and Ingraham report is even more pronounced with hockey gloves). But even with such large datasets, ocean currents are extremely variable, making specific object paths hard to predict. When I interviewed Jim Ingraham at his home on Whidbey Island just north of Seattle in 2012, a sea-worn blue turtle Floatee toy lost at sea some twenty years earlier was perched on his computer. Ingraham played a colorful computer animation of the paths of the bath toys migrating around the Pacific, some washing up on coastlines, others making laps with the gyre currents. "Anything could happen out there," he mused, pointing to the middle of the ocean northeast of the Hawaiian Islands where toy trajectories tangled into messy knots: the predicted accumulation zone that Ebbesmeyer had dubbed the Great Pacific Garbage Patch.

As global ocean current calculations meet particle tracking models and plastic samples from the sea, the garbage patch emerges in measured, and even bounded, form. Samples that I helped to collect at sea aboard the *Sea Dragon* in 2011, and later to process in marine ecology laboratories, quantified the contents from specific slices of the ocean surface into public datasets. As taken up by a variety of ocean researchers, these datasets joined many others in modeling shifting concentrations of plastic over time. Where the sample sites from the expedition were mapped as dots connected by the linear trajectory of the *Sea Dragon's* course, data visualizations of models tracking the garbage patch as a whole can have a tendency to carve a sea of fluctuating plastic saturations into bounded, even landlike forms. For example, take the image from a much-cited 2018 report in *Nature* titled "Evidence That the Great Pacific Garbage Patch Is Rapidly Accumulating Plastic"[47] (figure 1.4). It is also currently (as of 2023) the featured image on the Great Pacific Garbage Patch Wikipedia page. With solid and dotted outlines, the visualization brings the garbage patch into being as a discrete entity.

This is an "Alexander's eye" view drawn from above of the modeled concentration of plastic per unit of sea surface predicted for a particular point in time (this is not a satellite image!). A swirling continuum of comingled saturations of plastic and seawater is mapped into distinct classes of mass per square kilometer, color-coded, delineated, and named. With colors evoking a terrestrial topography, one cannot help but see "the modeled mass concentration" as deep water (dark blue), shallows (light blue), sandy

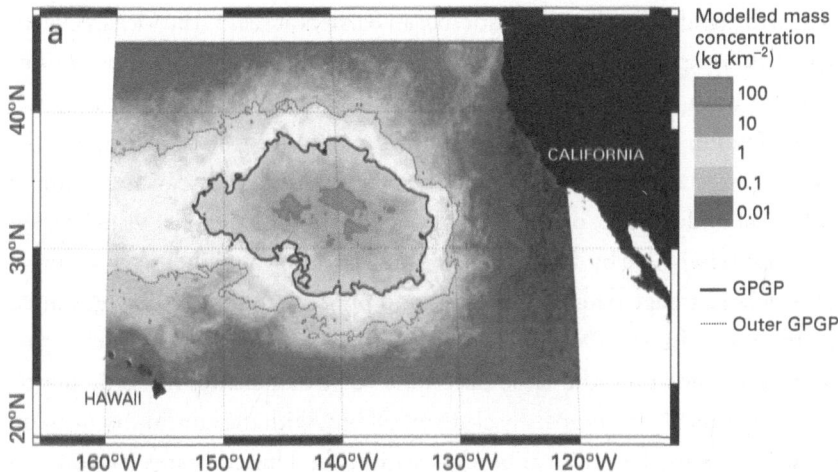

Figure 1.4

Great Pacific Garbage Patch Model, August 2015. The densest plastic concentrations are coded red in the center, fading into yellow and then blue. By L. Lebreton, B. Slat, F. Ferrari, B. Sainte-Rose, J. Aitken, R. Marthouse, S. Hajbane, S. Cunsolo, A. Schwarz, A. Levivier, K. Noble, P. Debeljak, H. Maral, R. Schoeneich-Argent, R. Brambini, and J. Reisser, https://www.ncbi.nlm .nih.gov/pmc/articles/PMC5864935/pdf/41598_2018_Article_22939.pdf, CC BY 4.0, https:// commons.wikimedia.org/w/index.php?curid=107377724.

beaches (yellow), and, circumscribed with a black border, a decidedly earthy (orange) form with several peaks (red). The legend explains the solid and dotted lines marking the "GPGP" (Great Pacific Garbage Patch) and "outer GPGP." From an STS perspective, this is an especially telling example of the performativity of classification, where the ontological entanglements of categories claiming to "describe" at the same time work to bring a garbage patch into existence.[48] But as an ocean scholar deeply conditioned by Western culture's terrocentric land/water divides, I also read these lines as the equivalent boundaries of geopolitical landform and accompanying exclusive economic zone (EEZ) that extends sovereign rights to natural resources for two hundred nautical miles offshore. From the perspective of these EEZ borderwaters, rather than borderlands, the United States has more sovereign ocean than sovereign land.[49]

This modeled garbage patch was funded by none other than The Ocean Cleanup with their artificial coastline, an elemental separation project that enacts the land/water line as plastic/ocean barriers at sea.[50] The Ocean Clean-up's striking appeal to landforms as "technology to extract plastic pollution" assumes both the possibility and desirability of rigid elemental categories of belonging; "extracting" plastic from the ocean depends heavily on the ability to see and maintain divides between plastic and seawater, human pollution and pristine ocean. In a stylized version of the data visualization circulated elsewhere by The Ocean Cleanup, shading and lines seem to imply a bounded garbage patch on one side and pure ocean that can be reclaimed on the other. In The Ocean Cleanup alternate version of the modeled garbage patch, however, the same deep blue color demarcates not only plastic accumulation from supposedly clean ocean but, with the continuity of color, similarly marks the political borders within the United States (states) and Canada (provinces and territories) but excludes Mexico. While I am assuming this was a design choice motivated by the desire for a contemporary brand aesthetic befitting a youthful tech-advancing nonprofit, this accidental equivalence is especially telling: The elemental boundaries of plastic/ocean, land/water, and the form of the garbage patch are betrayed as achievements as political as the outlines of Canada, the United States, and Mexico. The claim to be "ridding" the ocean of plastic is a similarly political achievement. Blue ocean in the models codes concentrations below thresholds that define pollution and harm rather than the complete absence of plastic. Such a model perpetuates what Max Liboiron calls pollution as colonialism: a sense of entitlement to use the ocean as a sink for storing waste by permitting certain kinds of plastic presences in the Pacific.[51] These models simultaneously double down on elemental divides where plastic belongs on land, not at sea.

WASTE(IS)LANDING THE PACIFIC

Like waterlines, rivers, and modeled garbage patches, islands too are acts of design, science, empire, and the imposition of ways of seeing. Emergent from these imbricated practices, a plastic trash island is an amalgamation of long-standing colonial practices of islanding and wastelanding in the Pacific, not a shocking anomaly of human products accumulating in an "empty" ocean "far" from land. Making this argument requires river scholarship be further brought into conversation with ocean scholarship, as the affinities between critiques of colonial water borders exceed both the line and its strangling boundedness to enable new multiplicities. I model the productive intersections of analysis that unite salty oceanic and fresher river scholarship, and critiques that connect island and continental settler colonialisms. While Dilip da Cunha and Irene Klaver write in detail of rivers, their analyses have many affinities with ocean and island scholarship, pushing back on Western colonial visions of islands that wield the line of land/water as a tool of power. Where river scholars are especially good at attending to lines, ocean scholars excel at unsettling boundedness. As Pacific Island anthropologist Alexander Mawyer argues, for many Pacific Islanders, "in an oceanic worldview the edge of the water is not a frontier at all."[52] Yet, the United Nations Convention on the Law of the Sea (UNCLOS), in defining an island as "a naturally formed area of land, surrounded by water, which is above water at high tide," solidifies with dry legal grounding assumptions of islands as preexisting, bounded entities circumscribed by the line of land/water.[53]

Island studies, and especially Indigenous Pacific scholars, have long worked to unsettle Western ontologies that naturalize islands as bounded and distant, as both separated from and by the sea. Following I-Kiribati and African American scholar Teresia Teaiwa's poetic invocation to "make island a verb," island*ing*, as taken up by cultural geographers Godfrey Baldacchino and Eric Clark, counters assumptions of natural geographic forms by emphasizing the ways they are actively produced as such.[54] I draw on the concept of "islanding" in two converging senses: first, in the performative ontological sense of da Cunha's line, as a set of practices that bring islands into being as a particular kind of (not so) natural geographic form—for example, what Epili Hau'ofa calls "islands in the sea," or Elizabeth DeLoughrey calls "the myth of the isolate," specks of dry land dwarfed by a vast uninhabited sea.[55] Such islands are places made distant and bounded, rendered ideal laboratories and testbeds for modern scientific and military endeavors. They are islands of decidedly western, colonial form. Second, I understand islanding more broadly, as the obsession with boundedness and discrete entities in Western culture. Islands are not only foundational units of modern geography's landscapes made distinct from continents but share with the disembodied elements of modern chemistry, processes of becoming and becoming known through institutional and epistemic violence.[56] Such modern building blocks, as Michelle Murphy urges, must be understood as relationships that include structural injustices.[57] Here islanding isn't just about making visible practices that enact the line; rather, islanding becomes a basis for challenging colonial, racial, and extractive violences that modern forms of boundedness enable.

In positioning the garbage patch trash island within this legacy, I am putting islanding into conversation with its continental cousin: wastelanding. As defined by Tracy Voyles, wastelanding is a technique of the settler colonial frontier that deems already inhabited lands "undesirable, unproductive, or unappealing" and the devastating practices that bring them into being as such, with all the racial violence that implies.[58] Writing of Diné dispossession in the American Southwest, Voyles describes the forced removal of Indigenous peoples and the subsequent destruction of centuries-old farmlands and orchards to render them into an "empty desert" fit only for extraction and pollution, for filling with mine tailings and radioactive waste. In his account of *Military Waste*, anthropologist Josh Reno draws on Voyles to argue that US imperialism in the Pacific has been made possible by Pacific islands that were "wastelanded, constructed to be heterotopias outside the limits of normal territorial democracy and capitalism, an illiberal exception to liberal freedoms."[59] By tracing expansions of the Western elemental line of land/water, I explore how wastelanding as a tool of US imperial power travels to aqueous oceanic spaces more specifically. Practices of islanding constitute an "empty" ocean that can subsequently be filled with waste, while simultaneously constituting plastic as matter that can be consolidated into new landforms.

Both wasteland*ing* and island*ing* draw focus to ongoing, performative processes by turning nouns to verbs. In doing so, both challenge what are assumed to be preexisting geographic conditions or elemental forms, empty land and bounded isolates, respectively, by focusing on the often-violent practices that bring them into being as such. Where wastelanding helps explain how oceanic spaces become full of pollution, islanding helps makes sense of why that waste insists on emerging in particular forms. Both processes are caught up with the rendering of inhabited spaces into empty wilderness, as *terra, aqua,* and *mare nullius.* For historian John Gillis, it is with Western forms of islands that the ocean itself becomes empty: "It was the Europeans and Americans who, when they entered the Pacific, introduced the concept of insularity, creating boundaries and isolating one island from another, turning the sea into empty space."[60] In his seminal periodization of the relations between modern economy and ocean space, Philip Steinberg systematically traces this transition from an ocean of danger into a surface to be crossed. For fifteenth- to seventeenth-century Europeans, the ocean was "a terrifying wild" drawn in vivid detail on maps: oceans of monsters and islands, both real and imagined, and a sea surface textured with waves. The modern European view of oceans shifted radically in the mid-seventeenth century, wilderness tamed by an empty grid. It is with industrial capitalism that the ocean is "drawn as a blue, formless expanse . . . the sea had become devoid of any substance whatsoever."[61] An ocean wastelanded, however, is not just empty but something unclaimed that can be expanded into. A sea "beyond possession" remained "filled with untamed and untamable elements," must come to possess qualities of *terra nullius* that Europeans had long reserved for the superior element of land alone. The Pacific as a "new world" manifests not with the first European explorations but later as an Enlightenment project. It is with the imposition of land-based forms of fixity that the ocean shifts from beyond the frontier of the enclosable, to itself become a frontier for new kinds of extractible and even inhabitable possibility.[62]

Frontiers, especially as caught up with Western advances into wilderness, conjure a whole constellation of territorial power relations, of settler colonial conquering and scientific and technological development tangled together in a seemingly unstoppable landscape of progress. It is only as part of a land- and continent-centric worldview that the Pacific Ocean appears so vast, remote, and empty. The Latin term *vastus* also roots the term for waste.[63] Building from thinking about waste(is)landing with Josh Reno, two place-making convergences of waste, islands, and colonialism in the Pacific are especially emblematic of the lineages grounding the synthetic frontiers of the garbage patch: islands deep with guano, nutrient-rich seabird droppings energizing the America frontier beyond continental coastlines; and islands as nuclear testbeds, controlled atomic sites (ir)radiating the New Frontiers of science and technology.

For the American frontier specifically, western is a divine compass declination set by Manifest Destiny. The Pacific Ocean is the cardinal direction of progress, setting a "natural" limit of a coastline to be reached and then exceeded.[64] The further from Europe, the more American you become. When nineteenth-century historian Frederick Jackson Turner delivered his hugely influential frontier thesis in 1893, the American frontier was at a literal turning point. Railways had reached the Pacific, settlers were backtracking to fill in remaining gaps, and the fragmentation of the moving boundary seemed to signal the vanishing of the frontier.[65] It is no accident that in this moment the first national parks were established, preserving wilderness areas where the privileged could continue to act out American fantasies at the edges of settlement.[66] Such distinctly white spaces and narratives conceive American freedom, environment, and environmentalism with Indigenous dispossession and enduring racial violence.[67] For Turner, this (unmarked white) American energy was so inherently expansive, that even the continental limit—the line of land/water—of the Pacific coast would be exceeded: "He would be a rash prophet who should assert that the expansive character of American life has now entirely ceased. Movement has been its dominant fact, and, unless this training has no effect upon a people, the American energy will continually demand a wider field for its exercise."[68] While Turner has been much critiqued for both historical inaccuracies and for championing colonialism, his thesis perpetuates the dominating logics of American empire.

The United States had already been staking claims to the "wider field" of the Pacific. In 1856, the Guano Island Act marked not only the beginnings of US empire beyond continental limits but also shows how wastelanding has long been foundational to island possession.[69] The act permitted Americans to occupy "unclaimed" islands to extract seabird droppings as cheap fertilizer for replenishing the depleted soils of industrial agriculture. Multiple meanings of waste—as the presence of guano and as the absence of productive use justifying occupation—were deployed with liberty. Islands claimed under the act included Palmyra, one of the Line Islands just north of the equator in the Pacific, which was later revealed to contain no guano at all.[70] On Pacific and Caribbean islands that did have guano, the labor conditions for Chinese workers were horrific, inciting riots and evoking comparisons to slavery by Frederick Douglass.[71] These island relations share many parallels with the very nineteenth-century European imperialism from which westward movement was meant to liberate itself, including associated "ways to take control over territory while avoiding many of the responsibilities that sovereignty implies."[72] Frontier movement is not only the expansive push of the boundary line but one that simultaneously cuts relations to intentionally limit obligations. Guano island settlement was supposed to be temporary—just until the waste ran out. Yet, when bird droppings were rendered obsolete by petrochemical fertilizers, the United States maintained island occupation: Pacific islands claimed as waste, including Midway Atoll, became military bases for protecting global flows of oil. Doing so simultaneously created legal precedent for decoupling settlement from sovereignty. Reno argues that guano islands were able to set such precedent, not merely through the (alleged) presences of extractable waste but because of how they were wastelanded, actively cut out of existing relations to place.[73] With guano islands, the American frontier emerges not merely as unfettered expansion beyond the "natural" limits of the line marking the Pacific coast but as entitlement to land with the clear drawing of boundaries that sever state obligations to specific peoples and places.

In 1893, the very same year Turner was raving about American frontier energy, the United States illegally occupied the Nation of Hawai'i—a formally recognized independent nation with many international treaties. Queen Lili'uokalani chose to yield, hopefully temporarily, to avoid immediate violence to her people.[74] Not even two decades later, a new national park was proposed for the island of Hawai'i, with a settler geo-logic that rendered the meeting of lava flows and the sea as the "most perfect" site for producing Western scientific knowledge about the formation of new rock that was generalizable to the earth as a whole. As historian of science Ashanti Shih compellingly narrates, the justification for establishing Hawai'i National Park connects American frontier relations to the preservation of "natural" spaces not only for recreational fantasies but as laboratories for Western science.[75] Frontier relations again displace Indigenous peoples and place-based ways of knowing, if always incompletely, with the violence of universalizing epistemologies.

Since World War II, oceans and outer space have been prominently positioned as frontiers in an explicit mapping of Turner's thesis onto the cutting edges of science and technology. In 1960, John F. Kennedy, having just won presidential candidacy at the Democratic National Convention in Los Angeles, declared in his victory speech the "New Frontiers" of challenges to be conquered by scientific and technological advances.[76] Calling attention to his embodied position standing at the Pacific coastline, "facing west on what was once the last frontier," Kennedy positions the past "behind us," confirming westward as the direction of progress, and the American frontier as the future:

> We stand today on the edge of a New Frontier—the frontier of the 1960s, the frontier of unknown opportunities and perils, the frontier of unfilled hopes and unfilled threats. . . . Beyond that frontier are uncharted areas of science and space, unsolved problems of peace and war, unconquered problems of ignorance and prejudice, unanswered questions of poverty and surplus. . . . Can we carry through in an age where we will witness not only new breakthroughs in weapons of destruction, but also a race for mastery of the sky and the rain, the ocean and the tides, the far side of space, and the inside of men's minds?[77]

Kennedy mobilizes the continental land/water line of the Pacific coast as it both directly materializes and more broadly symbolizes the assumed natural limits of American (and human) power. At the same time, he repositions the very limits of land as something to be transcended with technological innovation: Oceans, space, and problems themselves become new territory for expansion demanding new forms of elemental "mastery." With such innovations, oceanic depths and distant space alike become sites for expansion through exploration, resource extraction, and even potential habitation.[78]

In the Pacific, the New Frontiers of science and elemental American power reshape waste, islands, and colonialism with a distinctly atomic energy. Elizabeth DeLoughrey describes these relations as a new form of nuclear colonialism, with the United States claiming of Micronesia and surrounding waters alone tripling national territory. Just as national borders were expanding into oceanic spaces, nuclear experimentation was rendered permissible by promises of physical containment that produced islands as distant isolates in an empty sea.[79] Most infamously in the Marshall Islands, oceanic spaces and their inhabitants alike were made distant and supposedly isolated testing "grounds" for atomic military science and technology. Marshall Islanders were sometimes relocated for these tests but, at other times, were intentionally exposed to radiation and denied treatment to facilitate Western knowledge production about the effects of radiation exposure.[80] The line of land/water that constituted islands as isolated from continents enabled control/test group separations. These experimental conditions were propagated, in part, by nascent American ecosystems science funded by the Atomic Energy Commission itself.[81] Violently exploited as opportunities for research, Pacific islands and their inhabitants were waste(is)landed by nuclear weapons testing, land blasted into ruins, and racialized bodies subjected to radiation experiments without consent. As guano island occupation endures beyond the resource value of waste itself, so too does the active emptying of populated islands as temporary relocation for weapons testing become permanent displacement as flourishing island communities are made waste.

THE GARBAGE PATCH AS SYNTHETIC FRONTIER

As plastic pollution concentrates with currents and washes up on island shores, an ocean defined by the line continues to be treated as a place to extract and store waste. Plastic perpetuates relations of waste colonialism where European colonizers treat Pacific Island spaces as a dumping ground.[82] But with the garbage patch, waste(is)landing practices also converge in new ways: Land is not simply made discrete, possessable, and distant (islanding), or made empty and waiting to be filled with waste (wastelanding), but islands are made *with* and *of* waste. The frontiers of the garbage patch are not synthetic merely because they are made of plastic but because they are place-making expansions in the most elemental terms: The line of land/water, plastic/ocean constitutes islands even where there are none. With no recourse to being natural, the garbage patch instigates a much broader reckoning with on exactly whose elemental terms colonial islands and environmental crises together have been and continue to be made.

Describing the intense territorial reconfigurations of oceanic spaces over the last three decades, a group of Pacific studies scholars has drawn attention to a new line of tension.[83] They contend that contemporary expansions into ocean spaces through exploitation, control, and conservation constitute a distinctly Pacific frontier: a boundary they conceptualize between ocean grabbing and ocean communing. Where the interests of nation-states and corporations are closely aligned in staking territorial claims, they are also at odds when it comes to contests between an ocean of untapped resources to be exploited and an ocean wilderness to be preserved. Either way, the ocean is always already the void of Western ways of knowing violently enacted against situated Pacific Island epistemologies. As Fache et al. write, "the ideologies and the discourses of the frontier are inherently one-sided; they are based on the argument of an (imagined) institutional vacuum, which denies Pacific Islanders' historicized views of their sea/oceanscapes."[84] The ocean made vast, made empty, made wilderness, is made possessable by the imposition of techniques for colonizing land that further enclose aqueous spaces as state territory and private property: "In the Pacific, frontier entrepreneurs indeed create forms of spatiality that appear as projections to marine spaces of land-based devices, such as national boundary making, spatial planning, and protected areas."[85] The national parks model, for example, extends beyond dry land as marine protected areas, the metamorphosis, in part, of a much longer story of American frontier relations becoming oceanic.[86] Where military-industrial violence and modern science made waste, continued Pacific island occupation becomes justified by appeals to protect wilderness from the repatriation of Indigenous populations.[87]

By synthetic frontier, I emphasize how terracentric relations are not merely projected onto oceanic fluidities, but rather, practices of separation and containment work to physically rend land from sea. When plastic matter refuses to respect modern boundaries, whether elemental divides or nation-state responsibilities, it is the solidity of land itself that demands improvement—further controlling seascapes instead of adapting with place.

As synthetic frontiers, dominant solutions for plastic pollution give rise to artificial coastlines that further imposing land-making cuts of elemental control, whether through extractive cleanup, engineered barriers, or the production of new islands for settlement. The line cuts ever more landforms that can be further explored, extracted, and inhabited: the garbage patch as "a *terra nullius* literally expanded by those who have contributed most to rising sea level and plastic pollution."[88] Problematic inheritances of a global ocean plastic crisis bequeathed by previous generations animate quests to convince audiences that extracting plastic from the sea is not only moral but also profitable. The Ocean Cleanup's "boy genius" Boyan Slat, whose organization won the Ocean Hero Award in 2021, is the epitome of what Matie Zubiaurre calls the "trash hero." Solving ocean plastic pollution becomes yet another narrative "where the main character is male, a landfill invariably becomes a site of adventure and self-discovery," and "trash becomes an excuse for climbing mountains and overseeing a landscape as a metaphor for (male) conquest and appropriation."[89] Contemporary crises become new substrates for further domination.

As envisioned even more explicitly in fiction, ocean plastic pollution keeps insisting on becoming new land. The three-volume graphic novel series *Great Pacific*, written by Joe Harris and illustrated by Martín Morazzo, makes especially visible how western science and technology keep producing synthetic frontiers.[90] Packed with action, plastic trash, and seemingly every Pacific trope imaginable, *Great Pacific* details the exploits of Chas, a youthful Texas-born heir to massive oil wealth, as he attempts to settle a garbage patch island in the Pacific. There are rogue nuclear weapons, crashed Russian satellites, US Navy standoffs, Japanese businessmen, UNCLOS objections to a garbage patch state on grounds it is not physically moored to the seafloor, pirates, and tentacular battles with massive cephalopods. An opening set of panels establishes the location with a view not only from above but via satellite vision of militarized control with Chas as its target (figure 1.5).

Zooming in from space to precise coordinates, the garbage patch emerges from a murky swirl in the North Pacific Ocean, as if the bright blue of the scientific visualizations had sludge spilled all over them. This plastic isn't just solid waste in the ocean; it is consolidated into land where Chas can plant a

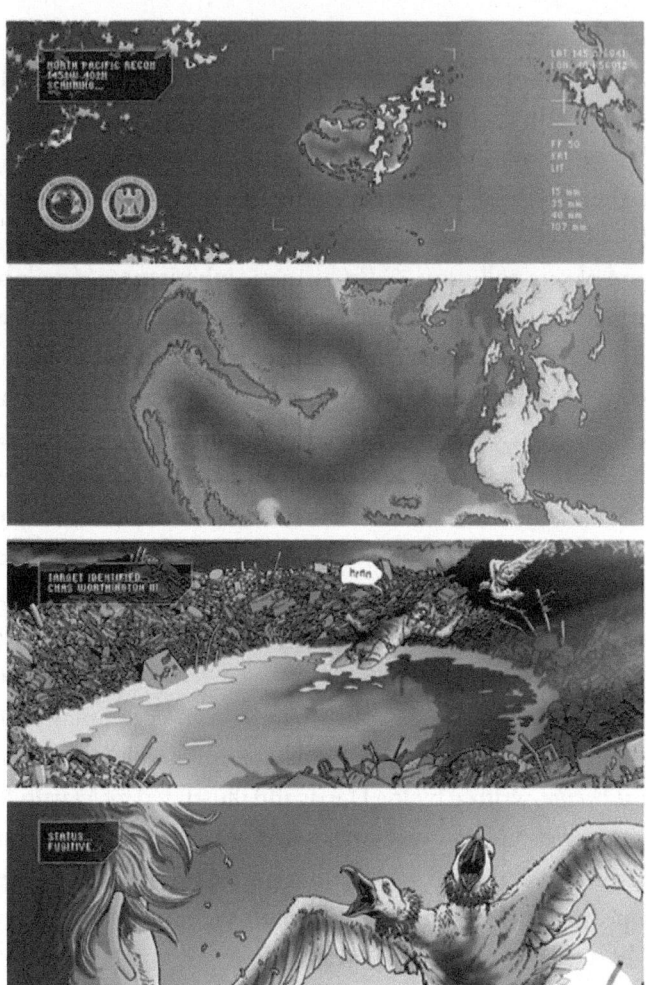

Figure 1.5
Zooming in on the garbage patch trash island from space in the opening panels of Joe Harris and Martín Morazzo's graphic novel series *Great Pacific*.

blue-green lone star flag and claim the garbage patch as "New Texas.'"[91] With technological ingenuity, plastic waste becomes grounds for the reenactment of the American frontier and its violent Pacific expansions. When Indigenous Pacific Islanders, who live past the frontier "to the west," arrive tattooed and loincloth-clad in outrigger canoes, Chas shoots the first one he sees. The series culminates with the revelation that his intentions had been environmentally motivated all along: Chas uses his "H.E.R.O" technology to transmogrify the garbage patch island into water vapor—elemental alchemy dissolving the paradoxes of plastic land back into sea—before bequeathing his wealth to fund ocean plastic research and a green energy transition.[92]

This would all read as sci-fi adventure—a heroic but hopefully ironic account of the attempted settlement of an oceanic garbage patch—if it weren't for its many direct parallels with contemporary responses to ocean plastic pollution, with The Ocean Cleanup and beyond. In chapter 2, the trash island vision manifests in full sincerity as "Recycled Island," a project spearheaded by Ramon Knoester of Rotterdam-based WHIM architecture, which envisions developing the garbage patch into a floating sustainable community with aquaculture, wind power, and dense urban areas explicitly modeled on the island of Hawai'i.[93] Similarly, Lenka Petráková's "The 8th Continent," awarded the 2020 Grand Prix award for architecture and innovation of the sea, images a plastic pollution cleanup and research station or even settlements.[94] While "interdisciplinary platforms" for addressing global problems might be necessary and admirable, with its long "barrier" booms and "collector" for extracting and recycling plastic waste at sea, The 8th Continent is the architectural embodiment of very widespread assumptions grounded in separation. Focusing on science, technology, and their Western elemental relations shows how terracentric forms of control are not merely imposed on seascapes; they are used to solve and reproduce crises—making new land and new plastic—in the same move.

(DIS)PLACEMENTS

The world where artificial coastlines are the *solution* to artificial coastlines is the same world where building a trash island is the solution to ocean plastic pollution: the world of synthetic frontiers. This chapter opened with a contrast between two plastic coastlines, Kamilo Beach and The Ocean Cleanup's interceptors, but it concludes with the assertion that they are part of the same set of processes that animate synthetic frontiers: land made discrete and distant and filled with plastic waste from afar; an ocean made empty, filled with plastic waste and molded into coastline or island form. When solutions impose the line of bounded elemental control, whether new land for settlement or extractive cleanup, they make distinctly synthetic frontiers. Moreover, the enduring power of the garbage patch capitalizes on expectations of a "pristine" ocean and the shock value of it being spectacularly trashed; human products in the Pacific are only surprising when the ocean is assumed to be an empty wilderness and when plastic matter is assumed to be endlessly controllable. Synthetic frontiers displace the frontlines of a plastic crisis to environmental elsewheres, "out there" in Hawaiian paradise and in the open ocean. Garbage patch island or not, plastic waste is already inextricably caught up with colonial island relations, as pollution washing up without consent is yet another instantiation of the ongoing occupation of Hawai'i, the land borders of the American frontier awash in plastic. Hawai'i has no plastic production facilities of its own.[95] If pollution is colonialism, then plastic coastlines are sites where the products of modern chemistry are instigating new landforms; Hawai'i's shorelines, garbage patches, plastic interceptor technologies, and trash islands alike are cut with modern science's synthetic frontiers. As the very emptying and distancing of the Pacific depends on a land/water divide, the territorial power animating synthetic frontiers is enacted not merely through environmental but distinctly elemental forms of control.

Ocean plastic pollution seems like a new, distinctly modern kind of pollution problem—in its heterogeneity, in the ways it acts like a fluid that keeps escaping our attempts to control it. But far too often it is addressed with the same old story: one that perpetuates human control over nature; people over stuff; white people over Brown; land over water; stories founded on the intersections of capitalism, colonialism, and racism. This is the story that promotes solutions that are part of the problem—extraction, growth, and yet more attempts to dominate—where even pollution itself is yet another opportunity. The Ocean Cleanup project deploys seemingly endless tropes of Pacific conquest in all seriousness. Slat boasts of "monumental" cleanup and promises new landforms. I do not think it is mere coincidence that fictional and actual proposals to settle trash island are emerging just as humans are confronting the limits of modern human control via climate change and rising sea levels. What better wasteland to bring into circuits of political and economic value than one made of actual trash?

In the summer of 2019, reports emerged that The Ocean Cleanup's first large-scale deployment was a failure after a portion of the floating device, including its "stabilizer frames," was found to be "drifting away."[96] Both islands and wastelands as ongoing processes can—and already are—being made otherwise. In the final chapter, "Synthetics: Placing Together/With," I take up work countering island isolates in the sea with seas of connection, replenishing scarcity with abundance. Pacific Islands as sites of contested epistemologies and ontologies bring forth other ways of water and other kinds of chemical relations, where the solution to plastic pollution is not more plastic coastlines, but the many lines of myriad oceanic relations. Yet cartographies that insist on the line of western land/water, ocean/plastic, and natural/artificial divides continue to produce bounded islands. The persistent line cuts synthetic frontiers with persistent designs: the garbage patch as trash island. By tracing elemental divides of plastic coastlines as they become forms of concern, I show how synthetic frontiers are so powerful that when their landforms cannot be found, they must then be brought into being. Technological-fix-privileging visions leap from the page as seemingly inevitable landforms arise from flows of plastic waste and seawater, perpetuating the very relations that led to an ocean plastic pollution crisis in the first place.

REORIENTATIONS

The three-day gap in my "field" notes from the *Sea Dragon* is marked with a desperate sideways scrawl: "Then we round the point." What I saw as a natural feature—a bit of land jutting out into the water, protecting us from the ocean—becomes a deeply embodied geography demanding all kinds of reorientations. The point where my land-situated self cedes ground to a visceral reckoning with the sea. The point where ambitious plans are tempered and expectations must be recalibrated. A mentor's seemingly simple ethnographic advice to "record everything," an imperative I had dutifully spelled out in capital letters on the first page of my notebook, becomes a physical impossibility. The regular inundations of seawater crashing over the bow confine cameras and notebooks below deck where the unfamiliar motion is even more nauseating. Multiple documentary film projects are on hold (one camera on board is rumored to be worth $100,000 and does not float). The number of projected scientific sampling sites is reduced from thirty-five to thirty, then again to twenty-five. The same wave action that is making people sick churns floating plastic below the surface, out of reach and out of sight. Battered by the rough waves, concerned about various projects, we are all anxious to get to the calmer seas that are mysteriously "just two more days away" for the fifth day running.

After the fateful first night, I never miss a watch shift, though I will never fully synch to our relentless schedule at sea. The *Sea Dragon* is not a fancy research vessel with support staff; it is a sailboat designed for adventure racing and chartered by a nonprofit.[1] Assigned to one of three watch teams, everyone is expected to contribute to the running of the boat around the clock, regardless of experience. Each day is divided into five watch shifts, tasked, in turn, with specific chores, along with taking the wheel to keep the *Sea Dragon* pointed in the right direction. The autopilot is out of order, or so we are told. I failed to foresee the implications of the math, where the resulting syncopation of three teams against five shifts would deprive us of two out of three solid nights of sleep for the three-week duration of the voyage. As consolation, the pattern also produces a fortunate side effect that we call Sunday: having both daytime shifts off duty and a full eight hours to sleep through the night. Each evening at 6:00 p.m., everyone on board makes a mandatory appearance for dinner. This is an opportunity for the skipper to give a daily update on our course and the forecasted conditions ("just two more days . . ."). It is also, he later confides, an opportunity to silently count bodies, making certain no one has been lost overboard. Life on the boat begins to settle into a pattern, with the days melting together into rotations of watch duties punctuated by mealtimes.

Nearly all our waking hours, day and night, are spent together with our watch team, organized to distribute our respective sailing, research, and media skills. I have the fortune of being placed on the team led by Marcus Eriksen. A long-time plastic pollution activist, cofounder of the 5 Gyres organization, and former Algalita education coordinator, Marcus is the center of well-deserved attention on board. He has already been on seven previous expeditions through gyres around the world, among other adventures, and regales us with stories of dinosaur digs, motorcycle crashes, and that one time he sailed from California to Hawai'i on a raft made of plastic bottles (and ran out of food). Marcus's experience is complemented by a pair from the Seoul Broadcasting Corporation: Womchin Jin, a high-status director of photography, and Brandon Kim, his assistant and interpreter. They are aboard the *Sea Dragon* on assignment, filming for a Korean documentary, but may not have been fully informed of the conditions of participation. They seem rather surprised to learn that they too will be responsible for cleaning toilets, cooking meals, doing dishes, and waking up in the middle of the night to steer the boat, a predicament that is no doubt exacerbated by reports that their colleagues are filming on a fully staffed luxury cruise ship near Tahiti. The footage is slated for a documentary with the working title *Last Pacific*, a reminder that the Western media does not hold a monopoly on narrating the plight of an endangered, even vanishing, ocean.

Learning to move with the boat leaves marks on bodies and challenges senses, but slowly we acquire new skills needed for life at sea. We compare bruises, field-test every seasickness product available, and learn to distinguish by sound in the darkness whether or not a wave hitting the bow is strong enough to soak your face when holding the wheel at the stern. Maneuvering the off-kilter world, we develop a sense of who and what is about to slide where and which little nooks are safest for rest. And finally, with eyes no longer fixed strictly on the horizon in seasick desperation, we gaze intently into the deep blue water, searching for signs that we have arrived in the garbage patch. Everyone on board swears they know the trash island sensationalized by the media does not exist as such, but I suspect we all got on the boat with lingering hopes of finding it or, at the very least, something spectacular. But at this point, we are quietly wondering if we will see anything at all.

2 THE TRASH ISLAND THAT ISN'T THERE

"The Pacific Plastic Trash Island: A Growing Environmental Crisis" proclaims the 2024 essay title, courtesy of ChatGPT.[1] Just seconds earlier, I had prompted the chatbot to write an essay about the trash island in the Pacific Ocean, not at all certain that the generative AI would comply. ChatGPT is programmed to refuse many kinds of requests, including those that spread misinformation. Testing the system while developing an undergraduate assignment, my prompts arguing for a flat earth and similar conspiracies had been shut down with the explanation that the results would "not align with scientifically accepted facts" and would "go against overwhelming scientific evidence." Though my trash island query passes algorithmic scrutiny, the essay's contents contradict themselves at every turn: The garbage patch is mysteriously "a massive trash island" and "not a solid island" at the same time. With a follow-up question, the chatbot assures me that the trash island is "not a myth" and simultaneously claims it is the product of "common misconceptions." While none of the accompanying sources directly support the claim that the garbage patch is a massive trash island, I do have some good ideas about where this language might have come from.[2] Trained on enormous past datasets, ChatGPT is especially adept at recycling dominant narratives, and the headlines of even the most reputable news outlets have reported the presence of trash islands in the Pacific. *Time Magazine* described a "swirling mass of plastic debris twice the size of Texas," with human impact on the ocean so severe "you can literally see the result." The garbage patch has been crowned "The World's Largest Landfill" by *Discover* amid calls to recognize it as "the 8th continent." As reported by the *San Francisco Chronicle*, *ABC News*, and *Oprah*, among countless others, the garbage patch spans hundreds of miles, is one hundred meters deep, and weighs 3.5 million tons. It is, following the most recited descriptor, twice the size of Texas. Or, in all its regional variations, "as large as Central Europe," with a "footprint as large as France and Spain combined," even "twice the size of America." Any remaining doubt about the garbage patch's substantial form had to contend with a headline declaring "Afloat in the Ocean, Expanding Islands of Trash," with the confident authority of no less than the "Science" section of the *New York Times*.[3]

Despite ChatGPT's acquiescence and the plethora of headlines, the existence of a trash island does not appear to align at all with scientifically accepted facts about plastic in the Pacific or any other ocean. At the height of trash island news reports in the 2010s, many of the ocean scientists I interviewed voiced their frustration with dangerous public misconceptions about the form of the Great Pacific Garbage Patch: The island did not exist. Despite the proliferation of claims about its location and size, no one could find it. Not in satellite images. Not on Google Earth. And not after weeks at sea. The trash island was a myth. What was missing was rigorous academic science communicated faithfully to help reconcile island concerns with material reality. At the prestigious Scripps Institution of Oceanography (SIO) in La Jolla, marine biology graduate student Miriam Goldstein, who had enrolled to study local tide pools, became so caught up in the apparent mismatch between trash island headlines and scientific knowledge that she switched labs to carry out research in the garbage patch. With a successful grant proposal, Goldstein was appointed chief scientist on SIO's first research cruise dedicated to ocean plastic pollution, a rare honor for a student. At the same time, Goldstein remained committed to both doing and communicating science. Resolved to correct all the misconceptions swirling about, her prolific public scholarship from the time includes an interview titled "Lies You've Been Told About the Great Pacific Garbage Patch," and an entire section of her well-read blog tagged "DEBUNKERY."[4] There was finger-pointing, too, at activists for exaggeration and at journalists for spreading misinformation. Goldstein became one of many involved with ocean plastic pollution who readily pointed me toward particularly egregious misrepresentations, hoping that as an STS and communication scholar I would be equally outraged.

I was not outraged; I was intrigued. I suspected something far more nuanced than a lack of science literacy or nefarious attempts to generate attention were afoot and hoped that getting to the bottom of things would foster more mutual understanding instead of blame. Looking for another entry point, I turned to Charles Moore himself, named specifically as a source of too-solid descriptions, and founder of the nonprofit Algalita Marine Research and Education, where I had been volunteering full time. Moore graciously agreed to a formal interview aboard research vessel *Alguita,* a fifty-foot catamaran safely moored under bluebird skies on a Naples Island canal, in Long Beach, California. We had been talking about the history of plastic pollution science and activism and the emergence of the term "garbage patch." My next question, and what I so desperately wanted to know, was who was the first to call the garbage patch a trash island? To my surprise, Moore pointed to "foreign papers," specifically *Pravda*, and proceeded to describe a captivating image of a "Matterhorn-looking mountain," an artist's conception of a floating landfill at sea. Immediately after the interview, I raced to search media archives, elated when Moore's tip checked out— the earliest mention of a plastic trash island did appear to be in *Pravda Online*, February 24, 2004. The short article, "'Trash Island' discovered in the Pacific Ocean," takes its content in turn from an article in German *National Geographic* equivalent *Geo* that describes a "carpet" of plastic in the ocean.[5] Exactly how the carpet morphed into an island remains a mystery of English-German-Russian-English translation.

To my great disappointment, the Matterhorn-looking image, like the trash island itself, was nowhere to be found. All that remained was a pixelated thumbnail of a blurred mound adorned with yellow caution tape and surrounded by water.[6] Hoping that the image might be preserved in another format, I asked the University of California, San Diego library for help finding a paper copy of the article in Russian, with hopes of tracing down the origins of the trash island myth at last. But in an act of academic triage, I was redirected to the Scripps Institute of Oceanography librarian. Despite explaining that I was trying to track representations of the plastic trash island, the librarian quickly cautioned me not to trust *Pravda*'s story, providing a link to a "more authentic and scientific source" instead. The source was SIO's own ocean plastic research project whose chief scientist, Miriam Goldstein, had pointed me toward so many problematic trash island representations in the first place. My quest for linear origins had done the opposite of getting to the bottom of things: It had returned me right back where I had begun. Though no closer to the elusive image, I realized that I had become caught up in well-meaning attempts to erase myths with what are seen as the objectively better representations of science, leaving only incomplete traces in their wake. Individual sources were not going to explain trash island's emergence. I was asking the wrong questions.

The persistent power of ocean plastic pollution in trash island form vastly exceeds attempts to attribute responsibility for origins in some head-line whodunnit. Instead of disentangling myth from fact, this chapter brings them closer together, to continue asking why ocean plastic pollution adheres so tenaciously to island form. My approach, building on matters of concern in STS, is meant to *add* to the realities of ocean plastic pollution rather than reconcile them into a single form, dissolve them into uncertainty, or reduce them to power alone.[7] In doing so, I join feminist new materialist projects of "getting further involved with their material-semiotic becoming: the coming to matter and ongoing mattering of things," while honing in on powerful forms that emerge and endure through such processes.[8] Trash island embodies the place-making performativity of synthetic frontiers rather than a mismatch between representation and matter that is or is not already there.[9] As chapter 1 traced the western lines that cut substances into discrete elements, island forms are caught up with enduring colonial land/ water divides. Moreover, the Pacific Ocean's very emergence as a Western geopolitical entity is entangled with speculative landmasses that may or may not exist.[10] Europeans were first drawn to the Pacific not only in search of a faster route to spice-rich lands but also in search of a southern continent assumed necessary to counterbalance the weighty continents of the Northern Hemisphere.[11] The mystery island-continent sometimes appeared on maps as Terra Australia Incognita: unknown land of the south.[12] This missing-mass tenet of Western geographical knowledge held from at least the time of Ptolemy's earth-centric universe, through the Copernican sun-centric revolution and into the nineteenth century.[13] When it finally gave way, due in part to colonial mappings of land/water lines constituting coasts in the Pacific, Western fascination with the missing landmasses had morphed into equivalent enthusiasm for land in bounded, isolated island form.[14]

Like land and water, modern Western traditions have treated matter and meaning as distinct entities. Supposedly neutral descriptions of how much plastic is at sea and what it is doing there are expected to precede conversations about whether it constitutes a problem and what, if anything, should be done. For decades, however, STS scholars such as Karen Barad, Donna Haraway, Bruno Latour, and Maria Puig de la Bellacasa have been pushing against exactly such assumptions by refusing to compartmentalize epistemology from ontology from ethics.[15] "Matter and meaning are not separate elements," declares the bold opening statement of Karen Barad's STS trajectory-bending *Meeting the Universe Halfway.* Drawing on physics that seem all too fitting for plastic swirling in an irradiated Pacific, Barad insists that the bond between matter and meaning is stronger than the forces holding eponymously indivisible atoms together: "Matter and meaning cannot be dissociated, not by chemical processing, or centrifuge, or nuclear blast."[16] With the seemingly irresistible convergence of existence and relevance in the word "matter," Barad and her colleagues invoke at once the qualities of physical substance and deserving attention. Moreover, the insistence on matter*ing* brings focus to ongoing processes that enact "substance and significance" at the same time, rather than presupposing a "gap" between knowers and the world, or myth and reality.[17] Insisting instead on inextricable connectedness, these relational approaches challenge foundational Euro-American divisions that subtend the production of objective facts as both independent from and overriding culture, politics, feelings, myth, and place. As scholars like Zoe Todd and Vanessa Watts must correct again and again, these elements have never been separate for so many Indigenous thinkers, just as so many Pacific Islanders see currents of connection rather than an ocean of isolation.[18]

Delving further into how concern about ocean plastic pollution has become caught up with the allure of islands, I show that the form ocean plastic pollution takes at sea is inseparable from its status as a global ecological crisis.

In other words, the trash island is a form of concern: Powerful shapes, dimensions, lines, and designs are entangled with whether the phenomena matters, and what, if anything, should be done about it. Where matters of concern, as articulated by Bruno Latour, call for presenting facts as they emerge with many entangled relations, his approach tends to neglect marginalized humans, even as it reassembles neglected things.[19] As feminist STS scholars continue to insist, there is ongoing need, in the words of Puig de la Bellacasa, to "reveal power and oppressive relations in the assembling of concerns" to stay attuned to "which worlds are being maintained and at the expense of which others."[20] In the case of trash island, positioning the substance of scientific accuracy against significant misconceptions has a tendency to expand worlds of synthetic frontiers rather than fact-check them out of existence. By staying with emergent contours and designs, drawing attention to form shifts emphasis from materiality in general to "cutting together-apart," to keeping track of binary differences as colonizing logics that can be reconfigured.[21] Forms of concern trouble expansions of synthetic frontiers rather than getting stuck tracing boundless plastic entanglements of everything with everything.

Trash island, then, is one of the persistent forms of synthetic frontiers. And synthetic frontiers, as territorial expansions of extractive worlds, cannot matter-of-factly be communicated out of existence. The trash island emerges, disappears, and reappears with worlds made with myth as much as they are made with plastic, science, and colonialism. As ChatGPT succinctly demonstrated, expanding islands of trash that figured as the stuff of headlines have refused to easily give way to scientifically accurate descriptions. As counternarratives circulate, knowledge of the trash island's absence does not make the problem of the persistent island—and especially the persistent plastic—go away. For the architects of projects like Recycled Island and The 8th Continent, trash island's absence itself solidifies as a problem to be solved by *building* one. At the same time, plastic continues to escape scientific attempts to know and contain it on a global scale. By exploring the contours of what is missing and for who, this chapter shows how the very existence of an ocean plastic crisis is inextricably connected to the generative potential of trash island's presences and absences and the enduring elemental lineages that make them powerful.

With all the trash island headlines swirling about, it is not at all surprising that when architect Ramon Knoester first read about ocean plastic pollution in 2006, he envisioned exactly that: a plastic landmass in the middle of the Pacific. For Knoester, founder of the Rotterdam-based WHIM architecture firm, the prospect of this new floating island was an image he found "horrifying and attractive at the same time."[22] From his perspective in the Netherlands, brimming with the longstanding threat of flooding newly compounded by rising sea levels, the trash island was not only a crisis but an opportunity for jointly addressing the challenges of ocean pollution and climate change. Knoester established the nonprofit Recycled Island foundation to "effectively address the plastic debris issue and create a new flood-proof habitat."[23] Like The Ocean Cleanup in chapter 1, the initial plans for Recycled Island involved gathering up plastic waste in the Great Pacific Garbage Patch. But where The Ocean Cleanup ultimately strives to bring plastic back to land, the Recycled Island Foundation envisions making plastic itself into new ground in the middle of the ocean.

Recycled Island puts the *synthetic* in synthetic frontiers (figure 2.1). A sustainable society built on the salvaged discards of a climate-changing world, it is a grand reclamation and development project for the Great Pacific Garbage Patch. As first circulated in the mainstream media in 2010, plans for the project boasted a whole new plastic landform: ten thousand square kilometers (3,861 square miles) of floating ground capable of supporting five hundred thousand people. The project descriptions are steeped with an expansive rhetoric of creative potential; of seeking, creating, constructing; of paths to new possibilities and, ultimately, to new land: "The proposal has three main aims: Cleaning our oceans from a gigantic amount of plastic waste; Creating new land; And constructing a sustainable habitat. Recycled island seeks the possibilities to recycle the plastic waste on the spot and to recycle it into a floating entity," explained WHIM's website.[24] The accompanying artist visualization depicts an enticing rainbow of substances neatly sorted by color into distinct sections forming an island—red, green, yellow, orange, and purple. The bright colors leap out in contrast to the merged blue surfaces of shimmering sea surface and pastoral sky. A Laysan albatross, the

Figure 2.1
Artist's rendering of the initial Recycled Island concept. A Hawaiʻi-sized island subdivided into colorful wedges afloat in the middle of the North Pacific Ocean. Image courtesy of CLEAR RIVERS.

poster bird of ocean plastic pollution known for ingesting plastic found at sea, soars across the upper right corner, banking inward, solid landing place in sight.[25] This hint of movement is echoed in gentle waves breaking into an occasional whitecap in the water below. Sunbeams stream through the clouds to illuminate a bright plastic future in concert with an again sparkling sea, empty except for sailboats and island afloat where there was presumably once only trash.

Through ocean cleanup, Recycled Island enacts elemental separations of plastic from ocean and land from water to bring the garbage patch into being in a specific form. Recycling is a kind of alchemy that renders plastic saturations into new land surrounded by pristine-again ocean. Land that will, in an echo of the first European forays in the Pacific in search of a missing continent, "return more balance to the environment."[26] Reformed as modular floating plastic bricks, waste collected in the gyre is to be melded into floating foundations for further development in the name of sustainability. These foundations are not a metaphor or (mis)representation of the garbage patch but the material substrate of its future. For Knoester, Recycled

Island should be the "final destination" for seemingly endless flows of waste, and the floating island their logical form.[27] Strikingly, the model for Recycled Island is not just any island. It is explicitly the Big Island of Hawaiʻi, which Knoester explains lends both size and shape to Recycled Island. While I am unsure if this influence extends beyond the general outline, a deliberate arrangement of what is assumed to be plastic sorted into colorful wedges that map onto distinct land uses: geometric shapes within hinting at urban and agricultural zones. The resulting radiating pie-piece divisions seem to echo the standardized model of traditional Hawaiian ahupuaʻa land divisions (chapter 1). Closer consideration, however, reveals that Recycled Island is far too regularly cut by canals to be following water in its many meaningful instantiations. Too-straight lines radiate as if from mountains to sea but across a strangely flattened topography in colors that bear striking similarity to the stripes of Dutch tulip fields.[28] Moreover, the design appears to cordon off human habitation into dense urban spaces separate from food-producing ones. In a second image, the albatross becomes a tour guide through a metropolis, gliding a canal lined with skyscrapers where palm trees sprout directly from plastic walkways. The smooth, rounded edges and symmetry bespeak a Jetson's meets Monsanto midcentury retrofuture. The postwar promises of freedom and prosperity through consumption is delivered at last through recycling and sustainable technology.[29] The goal of self-sufficiency powered by solar and wave energy, with seaweed harvesting a source of compost and biofuel, also resonates with a sense of the island-as-isolate, as liberation from continental relations beyond Recycled Island's shores, even as they are quite literally molded from those very flows. The New Frontier promise of science and technology conquering the elements by ushering forth "the mastery of the sky and the rain, the ocean and the tides" finally puts plastic in its place once and for all.

Recycled Island conceives the products of modern science and technology as both the problem and the solution: synthetic plastic as pollution and as a lost resource to be brough back into capitalist circuits of production and value through the endless manipulation of matter. Plans for Recycled Island embody the entrepreneurial spirit of synthetic frontiers, where problems are always opportunities. WHIM makes this quite clear among the multiple benefits of making Recycled Island: "This will clean our Oceans intensely and it will change the character of the plastic waste from garbage to building material. The gathering of the plastic waste will become a lot more attractive."[30] Ocean plastic becomes a lost recourse, the gyre a potential site for making cleanup profitable. Recycled Island both cleans up the ocean and brings waste back into the realm of useful, valuable material, where value is anchored in making plastic matter landlike and habitable for humans.

There is, however, some confusion over the very practical matter of whether the whole project required constructing an island or simply developing one that was already there. Some media coverage describes the technical difficulties of plastic collection, given the size of the Pacific, noting that "it will take years to gather enough plastic before there's enough to melt together to form the gargantuan island."[31] But those commenting on these articles had other ideas about what Recycled Island entails: "I think to build on top of it would be a bad idea. It's just human to cover it up and forget about it but it's still there," wrote one, while others jumped to weigh in on what to do with the floating garbage island.[32] Did the trash island already exist? Could it just be towed to shore instead? Or would the island form itself need to be brought into being as the foundation for Recycled Island?

DISSOLUTION

That the existence of a Hawai'i-shaped or Texas-sized trash island was even a question is exactly the kind of conversation that was leaving many in the ocean science community exasperated. Where activism and media coverage had generated considerable public concern, ocean plastic pollution did not seem to register at all among academic scientific priorities. All the trash island talk seemed to be bringing about a deluge of impractical solutions from people concerned about the wrong things or, worse, fully imaginary ones. Goldstein's student office at SIO, cheerily adorned at the time with PhD comics and a poster of Barbie exclaiming "Plastic is great!," also featured a photo taken in front of the Capitol Building of a sign that reads: "What do we want? Evidence-based change! When do we want it? After peer review!" The extent to which ocean plastic constituted a crisis and what, if anything, should be done were questions best tackled after accurate description. As for many scientists, for those I worked with, accuracy implied descriptions that mapped directly onto empirical observations and sample collection at sea. Meticulous work in the field followed through to the lab and modeling resulted in matters of fact: claims about the world that are independent of human interests and feelings. Having an agenda, especially a political one, gets in the way of the skepticism necessary for good science. "If I'm not sure, then it's not plastic," Goldstein explained, while expressing concern that activist scientists seemed to always find plastic without fail.[33] The measured density of plastic particles on the ocean surface should not change in amount or form, no matter how worried you might be about plastic pollution. Following still dominant models of science communication, good representations convey scientific knowledge without distortion.[34] Levels of technical detail might need to be adjusted for less specialist audiences, but the shape of the problem should remain the same regardless. While practicing scientists often concede that such ideals are never perfectly achievable, any "gap" between scientific descriptions and public understanding is to be minimized as best as possible and should most definitely not be large enough for new continents to form. At the very least, plans to develop the garbage patch should not precede accurate knowledge about what is really there (or not).

Goldstein was not the only ocean scientist who altered the trajectory of their own research to quell the flames of myth by doing and faithfully communicating good science. Also among those resolved to get the bottom of trash island tale, was Dr. Angelicque White, now associate professor of biological oceanography at the University of Hawai'i, Mānoa. White's research focuses on marine microorganisms in the Pacific, such as *trichodesmium*, bacteria that glom together in spikey bundles forming colonies sometimes big enough to be visible to human eye and, in just the right conditions, from space. In 2008, White, then faculty at Oregon State University, had been getting a lot of calls. Not about ocean microbes as she had hoped, given her area of expertise, but about ocean plastic. White was perplexed. She has spent hundreds of days at sea, done over a thousand net tows collecting microorganisms from the surface of the Pacific, but had never once found plastic in her samples. Set on investigating further, White headed to the garbage patch as a research scientist with SUPER (Survey of Underwater Plastic and Ecosystem Response), the first research cruise to study the impact of plastic on microbial communities in the garbage patch.[35] While more charismatic turtles, birds, and fish had been the stars of ocean plastic conversations, microscopic animals and plants constitute 98 percent of the ocean's biomass.[36] White, with all her ocean experience and academic skepticism, was still surprised at how little plastic was out there. It was not visible on the surface, and even with a long tow of the sample net (unlike the much shorter ones she did for her usual microbe research), the results were relatively low concentrations. She wondered, as many have since, if maybe they "were not in The Patch."[37] Scrutinizing the existing data, trying to make sense of how an island grew seemingly from nothing, White estimated that if all the pieces of plastic floating on the ocean were scrunched together, it would be merely 1.4 percent the size of Texas (a calculation that she ran by Goldstein before making public). Though maybe, she conceded, more people would listen to scientists if they took to using Texas as a standard unit of measurement.

The results of White's ocean plastic research were shared widely via a 2011 University of Oregon press release provocatively titled "Oceanic 'Garbage Patch' Not Nearly as Big as Portrayed in the Media."[38] Quoting White directly, the press release positions measured scientific calculation against media sensationalism: "There is no doubt that the amount of plastic in the world's oceans is troubling, but this kind of exaggeration undermines the credibility of scientists. . . . We have data that allow us to make reasonable estimates; we don't need the hyperbole."[39] Through the report, White systematically disputes the major facts circulating about the garbage patch, arguing that it is at most only a fraction the size of Texas, that there is no evidence it is growing exponentially, and that it is not in any way an island, especially not the kind that could be seen from space. The press release leaves little doubt about the conclusions, giving White the final word: "If there is a takeaway message, it's that we should consider it good news that the 'garbage patch' doesn't seem to be as bad as advertised."

In my own work, I encountered references to the press release over and over again, from Goldstein, at the Algalita office, through systematic media analysis, with a common theme: White's voice was being invoked as a credible source of dissent supporting claims that the size, shape, and severity of garbage patch problems were vastly overblown. Given White's apparent hard science verging on pro-plastic stance, I arranged an interview with some trepidation. Who was she? I wondered, and what interest did she have in contesting the issue? Surely, she must be some variation of the climate change denier conservative, possibly with corporate ties (yes, I had been reading Oreskes and Conway's *Merchants of Doubt*).[40] Instead, I found a passionate researcher and teacher, with a genuine sense of care for the sea grounded in a deep respect for Western scientific methods. White, it turns out, was not sponsored by the plastics industry or any other nefarious source. She was not funded for her plastic research at all. All the attention the plastic press release was garnering was only adding to her dismay that public priorities were in the wrong places.

If White was by no means the voice of the plastics industry, she was the voice of reason, committed to undoing irrational public responses with rational science: asking questions, thinking critically, and evaluating the evidence. She patiently walked me through a slide presentation titled "Hyperbole and the North Pacific Plastic Patch." In the cover image, a photograph of the Pacific, there is the familiar two hues of blue meeting at the horizon, ripples on the deep blue sea surface below, and a scattering of fluffy clouds in the light blue sky above. Conspicuously absent is not only the trash island but visible plastic of any size or shape. The green metal winch system used to raise and lower sampling equipment from the large research vessel occupies the bottom-left corner of the image, imposing the stamp of institutional science on the otherwise empty-seeming seascape. A wake of white bubbles evokes movement toward the open ocean, as the mostly unseen ship passes out of the frame. "So, you ended up with very different results using the same data?" I asked. "The *same* results," she clarified. Following Algalita's own sampling methods, White's analysis disputes not the measured quantity of plastic in the ocean but rather the broader conclusions being communicated with that data. The material evidence does not support the meanings that have become attached to it after the fact. In person, too, White seems to agree that plastic problems are not so dire, arguing that there are still areas of the ocean largely unpolluted by plastic. When it comes to distributing scarce resources for research, plastic does not seem to rank at all among the greatest threats to the ocean. Yes, there is some plastic in the ocean, White concedes, but we cannot fix everything and need to weigh the risks.

Far more troubling for White is activist hyperbole and media hype. And they do more than create state-sized problems from dispersed fragments; they provide grounds for dismissing the credibility of nonprofits, media outlets, and scientists alike. Trying to reduce the amount of plastic in the world at any cost is what White calls "activism with blinders," a slippery slope to distortion, hyperbole, and lies.[41] She does not shy away from naming names, tracing the much repeated suspect claims—an ocean with "more plastic than plankton," a garbage patch "landfill" that is "twice the size of Texas" and weighs "3 million tons"—back to direct quotes by none other than Algalita founder, Captain Moore. Though members of Algalita do have experience at sea, they do not possess requisite skepticism and are "undermining the credibility of scientists." The implication is that concern on grounds of anything other than the best knowledge (read: accurate academic science) is detrimental to not only to the cause but also to scientific projects more generally.

On her website, White further clarified the implications of a lack of Texas-sized floating trash island: "This is not to say that the issue of plastic in the ocean should be dismissed; rather, the problem is more complex and enigmatic than that conveyed by the imagery of a cohesive patch spread out over a few remote locations." Yet despite the nuanced attention to detail, her claims get taken up to do just that: to dismiss the problem, and not just the problem of a trash island. In the majority of cases, the press release is simply reproduced with few, if any, alternations. This is common practice in science journalism in part because journalists must cover fields far broader than their areas of scientific expertise.[42] When modifications do happen, journalists tend to frame the report as either the latest word on the garbage patch or as yet more science discredited by other science. But usually it is only the headline that changes noticeably: "'Great Garbage Patch' in the Pacific Ocean Not So Great Claim Scientists," in *The Telegraph* (Alleyne 2011); "Giant Floating Trash Pile Not So Big After All, Prof. Says" in the *Seattle Post Intelligencer* (Ho 2011); "Claims Island of Plastic Waste Twice the Size of Texas Is Floating in the Pacific Are 'False'" in the *Daily Mail* (2011).

The cumulative effect, that neither the garbage patch or trash island is "so great," diminishes plastic problems in general. A smaller garbage patch, one that is not so solid or visible, these articles convey, must be good news for the ocean. The absence of a trash island is equated with the absence of an ocean plastic problem.

Despite White's obvious skepticism about the extent of an ocean plastic problem, her encounter with the missing island and measured evidence only fuels the expansion of synthetic frontiers in the hands of others. The press release claims are very intentionally mobilized to defend the continued production and use of disposable plastic. In a provocative example, Marc Gunther, author of a business and sustainability feature on *Greenbiz* and regular contributor to *Fortune*, leverages White's measured scientific caution "in defense of the plastic bag."[43] Constructing a pro-plastic manifesto to counter a tide of plastic bag bans and taxes, Gunther juxtaposes White's expertise as an ocean scientist against alarmist claims about the problem, courtesy of Oprah (who had declared the garbage patch "the most shocking thing I've ever seen"):

> Whether Oprah has actually seen the garbage patch is anyone's guess. But Angelicque "Angel" White, an assistant professor of oceanography at Oregon State, participated in one of the few expeditions solely aimed at understanding the abundance of plastic debris in the Pacific. He [*sic*] says the claim that the "Great Garbage Patch" between California and Japan is twice the size of Texas is flat wrong.

White is not only a professor but an expedition participant through a scare-quoted "Great Garbage Patch." Gunther borrows her experience, backed with institutional position, as evidence that "plastic pollution of the oceans probably isn't as bad as you think." The question of how to care becomes a question of whether anyone should be concerned at all.

REFORMATION

On one of my first days volunteering at Algalita's Long Beach office back in 2012, I was handed a copy of a just-released issue of *Earth* magazine, a publication dedicated to chronicling "the science behind the headlines." The cover story was about the work of tracking plastic waste in the ocean, with Captain Moore and Algalita featured prominently among the first organizations to raise awareness about the problem. The author does an especially good job describing the nuances of plastic fragmenting and sinking, of the ever-moving currents, and the uneven distribution of floating materials. The article very explicitly states that "the garbage patches floating in the oceans are not solid, giant floating islands of trash," with accompanying images from Goldstein of microplastics scattered on the sea surface.[44] But as I read, a concerned-looking Algalita staff member pointed to short section halfway through, that begins "Not everyone is convinced that the Great Pacific Garbage Patch contains as much plastic as Moore and his team calculated." The few paragraphs that follow point to skepticism about the amount of plastic in the ocean, emphasizing the difficulty of producing accurate measurements and the lack of funding for doing so. The source is none other than Angelicque White, with her claims arranged as a neat counterpoint to an article that would otherwise be completely dedicated to the details of carrying out research on ocean plastic problems—problems that Algalita and its founder Moore had spent over a decade tirelessly working to bring to public attention.

Algalita might not have had the resources of a major research university, but the organization did share White's concern that their science was sound and their facts credible. They too were frustrated with the trash island form, and especially with the way it was impeding practical solutions to ocean plastic pollution. Algalita's community relations coordinator had been fielding seemingly endless emails and phone calls from all kinds of earnest people with plans for cleaning the ocean of plastic that so often assumed big islands of trash already out there. The kind of island where a boat could be tied up to start working on recycling the plastic right out of the sea. Or where polar bears might seek refuge from a melting arctic ice cap. Echoing White, these visions that were problematic because they did not map onto the shape of what was out there. As Algalita's coordinator explained, "They start building on this false impression . . . wasting time and energy on something where they can be devoting it to something else."[45] The concept of trash island creates extra work for those sharing knowledge about the ocean, as they must re-explain and push against solid ideas people bring with them.

Algalita members, unlike White, were adamant that even in the absence of a trash island, ocean plastic pollution still was a major problem. An island absence was not at all "good news" for the ocean. The Algalita team, like many others, found themselves addressing audiences that were now showing up not only with knowledge of the trash island but usually, because of it, with well-intentioned concerns that nonetheless needed reshaping. Surely it must be possible to reform concern generated by trash island, to channel it toward other kinds of responses and solutions rather than dissipating hard won attention to the issue entirely. Care for ocean plastic pollution should be in the right form, so that meaningful change would follow.

How to communicate that form with concerns intact, however, was proving to be a struggle. Almost everyone I have interviewed, whether working in nonprofit or academic worlds, expressed worries about the limitations of "garbage patch" as a descriptive term. It is not an accurate depiction of the marine debris problem in the North Pacific. Garbage patch is far too terrestrial, bounded, and solid; it too easily coalesces into floating landfills and trash islands. The language of scientific alternatives such as "concentrations of marine debris" (NOAA), however, was hardly the gripping stuff of headlines. Moreover, these alternatives tended to dilute responsibility for plastic problems. Moore had tried to replace garbage patch with "a swirling sewer" and even "a superhighway of trash" connecting two "trash cemeteries," with little success.[46] The island alternative that has gained the most traction is plastic soup. In his 2011 book, *Plastic Ocean*, separating trash island from plastic soup is among the first orders of business:

> Let it be said straight up that what we came upon was *not* a mountain of trash, an island of trash, a raft of trash or a swirling vortex of trash—all media-concocted embellishments of the truth. It would become known as the Great Pacific Garbage Patch a term that's had great utility but, again, suggests something other than what's out there. It was and is a thin plastic soup, a soup lightly seasoned with plastic flakes, bulked out here and there with "dumplings": buoys, net clumps, floats, crates, and other "macro debris."[47]

In place of trash island, Moore offers up the alternative and more aqueous form of plastic soup. At all kinds of Algalita events and presentations,

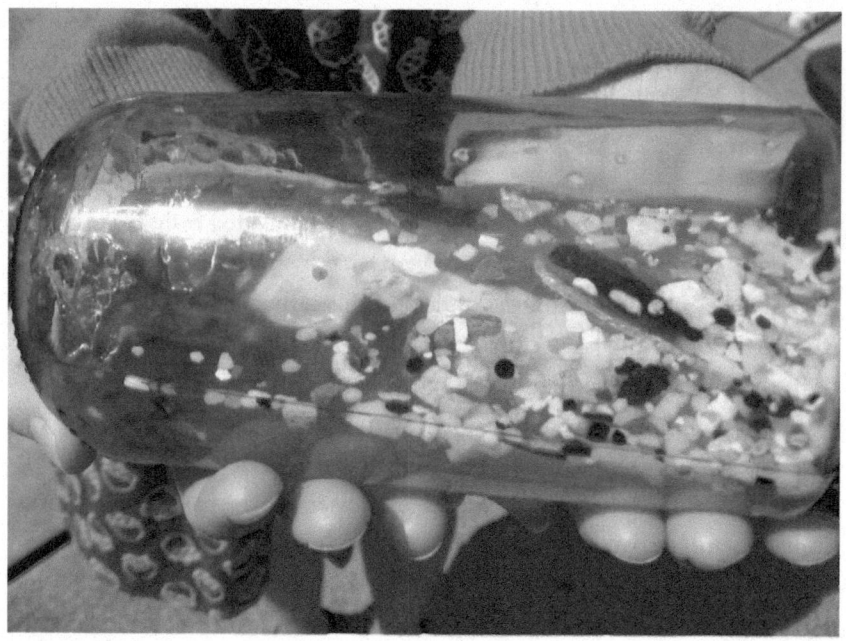

Figure 2.2
Displaying an education sample jar of "plastic soup" collected on the 2011 North Pacific expedition. A mangled black pen cap and many bead-shaped industrial nurdles are visible among the fragments. Photograph by the author.

plastic soup was described and shown to confront the trash island or other too solid versions of the problem. Glass jars filled with bits and pieces of colorful plastic suspended in murky liquid, presumably seawater, are held up and passed around as samples of the garbage patch (figure 2.2). My own jar, one I helped collect from the garbage patch, still sits on a shelf in my office. Upon closer inspection the plastic is mixed up with now-dead sea life, lantern fish, small crabs, and gloopy plankton. For Algalita's education coordinators, these jars are the key to informing the public: "The best way to get somebody to understand is to have a physical sample with you, to kind of like shake it up and show them that it's more of a soup."[48] These many activities coalesced around the task of rendering the island absent while maintaining the presence of plastic problems: shifting how the public understands the shape of the problem from trash island to plastic soup, while maintaining interest and support. The island may be absent, but plastic problems are definitely present.

Dishing up plastic soup, however, does not in itself convince the public that the island is not there. One metaphor does not smoothly replace the other in a linear progression toward objectively better representations of ocean plastic's "true" form. Rather, multiple forms of ocean plastic pollution circulate, overlapping, bumping into one another. Moore was describing an ocean of plastic soup and showing people samples of it in 2002, at least five years before stories about trash island proliferated. More than ten years later, there were still the many assumptions about the size and solidity of the garbage patch to contend with. Giving a presentation to a packed auditorium at the Aquarium of the Pacific in Long Beach in 2012, Moore reappropriated the twice-the-size-of-Texas claim. It is not a trash island twice the size of Texas, he clarified; rather, the area we were *sampling* is roughly twice the size of Texas. Then, after an hour-long presentation about microplastics, toxins, and synthetics all caught up in animal bodies, my eyes widened as I listened as audience members continued to ask completely serious questions about cleaning up "the island." There was a shared sense, from both academics and activists, that media coverage was improving, that public awareness of plastic soup was increasing. More reputable media outlets took up plastic soup or at least dropped the trash island talk. Still, in 2024, more than a decade later, and I find myself watching my own students' final class presentations complete with images of a trash island garbage patch, or in conversations about my research with reputable scholars who are sure there is in fact a trash island—they have seen pictures of it.

There is a tendency for plastic problems to disappear along with the trash island. But far more often, those working to make and share knowledge about ocean plastic pollution lament the tenacity of trash islands. You can get out the sample jars, show all kinds of images of microplastics, talk about toxins in seawater, about plastic fragments stuck in bodies; but where a plastic problem is present, the trash island will not go away, not even when serving up plastic soup. Moore's engagement with "trash island," whether in newspapers, online, or during my fieldwork with Algalita, all seemed to be tied to saying it does not exist. Or, at least, not yet: "Well, you know, it's been referred to as a 'trash island,' but really, what we're trying to do is stop it from becoming a trash island," Moore has clarified.[49] Trash island lingers in its many potentialities as threat and as opportunity. Despite all the work of correction, substitution, and reappropriation, trash island does not stay dissolved into an ocean of truth and microplastic saturations. The form of ocean plastic itself becomes a concern. Ocean plastic becomes so inextricable from island form, its presence looms, if not already existing, as something that could/would/will come to be.

ISLAND BUILDING

The fact that a plastic landform was not yet already there awaiting development did not deter Ramon Knoester from his Recycled Island plans. A trash island in the North Pacific was not a myth or misconception to be corrected, but rather something to be conceived. Trash island is both the problem and the solution: It is missing, but it can be built. WHIM's project plans with their vivid images captivated broad audiences and generated all kinds of responses. One *ABC News* headline makes Recycled Island sound like a fait accompli, announcing "Pacific Ocean to Receive Plastic Island," as if it is only a question of when. That said, most commentators are more cautious, labeling the project "ambitious," "bold," and perhaps even "impossible."[50] Goldstein, too, weighed in on Recycled Island, tirelessly upholding a reality of dispersed, microscopic fragments, of "thin soup," while lamenting in jest that a plastic island in the middle of the Pacific would have been a very convenient platform for her own research.[51] Much more seriously, however, she noted the incredible challenge of removing plastic without taking all the zooplankton with it—a move seemingly very much at odds with Recycled Island's sustainable dreams. Building floating trash islands, like disappearing them, was proving more difficult than expected as plans formulated on land encounter the challenges of the open seas. Knoester was well aware of the challenges:

> Ah yes, it is not easy because it has never been done before. The plastics are difficult to collect. The plastics are altered by the sun and ocean water. The Island should be very strong to withstand high waves. And this is all true. If it

was easy, then we would already have several Islands floating there. We have to make a big effort to make Recycled Island happen. Sometimes complexity is mistaken for something else.[52]

The mismatch between island vision and scattered microplastic fragments becomes a technical problem to be overcome. The missing island itself becomes a problem to be solved not with more accurate description but by building one—the island absence presents an opportunity for expanding synthetic frontiers. Vocal critiques of the project from the scientific community, which almost always focused on technical impracticalities, only seemed to further technical developments. A Recycled Island Kickstarter promotional video from 2012 gives considerable space to demonstrating techniques for island building rather than an accurate description of what is already there. The concept of floating trash islands survives despite attempts to dismiss it as impossible or to dilute it into plastic soup too difficult to strain. Matters of accuracy and credibility through good science are usurped by matters of feasibility and conviction. The one hurdle that seemed insurmountable was the problem of cost. The Kickstarter campaign generated a lot of attention but ultimately went unfunded. And gathering enough plastic from the gyre itself was proving too expensive.

In the years since, Knoester has focused his efforts much closer to home, trapping waste from urban waterways, including along the Maas River in the Netherlands. Here plastic is more concentrated and less degraded, and land is claimed not from Indigenous peoples but from the sea. Though still officially under the umbrella of the Recycled Island Foundation, Knoester has established the Clear Rivers project and rebranded accordingly. "For a plastic-free sea, we act at the source!" exclaims the website.[53] Yet core tenets of Recycled Island as synthetic frontier still hold: capturing plastic waste to reform it into a building material and using that to make floating land. These projects seem rather modest in comparison to the original vision, morphing from a giant ocean metropolis to a 140 m^2 (1,500 square foot) park, a bare fraction of the size. Beginning with smaller projects was precisely the point. Recycled Park, and then RE:Villa, a floating house, would demonstrate the feasibility of scaling up to Recycled Island. Recycled Park's concept images

Figure 2.3

Recycled Park floating on the Maas River in Rotterdam. The black plastic hexagons support flowering plants and a human seating area. Photograph courtesy of **CLEAR RIVERS**.

merge before and after images of the river to position a bright floating park surrounded by clean again water, with birds and fish frolicking in contrast with a river previously inhabited only by a stream of trash floating toward the sea.

The Clear Rivers website extends an invitation to come visit Recycled Park (figure 2.3). If you make your way to Rotterdam, in person or on Google Earth, to a particular side canal of the Maas River, there sits the Clear Rivers office in a sustainable floating house topped with solar panels. Alongside is a collection of hexagons, edges a bit too regular and straight, like the concrete banks of the canal and river, like the waterways of Recycled Island, or the boardgame tiles from *Settlers of Catan*. A few of the black plastic modules have benches for humans; the rest sprout green grass and flowers, not the microbial plants of the open ocean but a terrestrial view of a healthy ecosystem. Emblazoned on the plastic park itself is an explanation that reads: "The plastics are recycled and give new value to the river. From the plastics we construct floating platforms for a new green environment." Recycled Park might not be Hawai'i-sized, but it is indeed floating land built of reclaimed plastic waste.

Knoester is by no means alone with plans for ocean plastic that envision creating new landforms at sea. In 2020, Lenka Petráková's "The 8th Continent," was awarded the 2020 Grand Prix award for architecture and innovation of the sea.[54] As envisioned by the designer, The 8th Continent is a plastic pollution cleanup and research station intended for the Great Pacific Garbage Patch. Petráková sought design inspiration in the interactions between living organisms and their environments. She envisions the station as "dynamic architecture . . . influenced, transformed and organized by ecology systems."[55] The results are visually striking, like a cross between the Sydney Opera House and a hummingbird. In mesmerizing imagery of the station in motion, the sails become wings, buzzing harmoniously with the movements of the ocean. Inside the floating station, the imagined facilities include greenhouses for food production and water filtration; sorting and storage areas for collected ocean plastic; and research and education centers, all in self-sustaining, floating continental form. There is a lot to admire about this proposal: the fluid aesthetics of the design itself, the goal of changing *relationships* with the ocean, and the inclusion of an education center instead of limiting the focus to science and technology alone. There is also so much that resonates with the expansion of synthetic frontiers and their decidedly terrestrial, expansive ambitions. There is the obvious connection with The 8th Continent name but also with the vision of a self-sustaining floating landform and especially, the overall extractive aim of cleaning up ocean plastic and ultimately recycling waste into new products.

Elsewhere, powerful intersections of technological progress, environmental crisis, and capitalist expansion make themselves even more explicit in island form. "Long before humanity colonizes Mars, our children will live in floating nations," prophecies the voiceover in a *BBC Ideas* video. Floating architectures, as Melody Jue contends, reduce risks such as those associated with sea level rise to problems of geography—a shortage of dry land—further grounding capitalist reproduction in the imposition of western territorial relations.[56] Meanwhile, the Seasteading Institute (cofounded by champion of neoliberalism Milton Friedman's billionaire grandson) promises to build floating islands to establish a libertarian paradise, free from land and from state. "There is no space for experimenting with new societies," laments the institutes website.[57] While their initial Blue Frontiers project seems to have fizzled, the Institute's list of active projects reads like a catalogue of synthetic frontiers—hydrogen energy stations, single-family seasteads, private clubs, entire real estate ventures—all floating, all "environmentally restorative," all to be realized by "building on the ocean."[58]

WHAT IS MYTH/MIS-/MISSING

Reclamation schemes and fantastic new tech might seem absurd, but it is no random coincidence that ocean plastic in the Pacific keeps coming up as islands in particular: as an elemental form of colonialism and Western science. And keep coming up islands it does. Years after the initial *New York Times* "expanding-islands" headline,[59] with all the institutional science, with all the efforts at correction, Moore himself publishes an opinion piece—in the *New York Times*—that describes recent garbage patch encounters in terrestrial terms:

> Plastics of every description, from toothbrushes to tires to unidentifiable fragments too numerous to count floated past our marine research vessel *Alguita* for hundreds of miles without end. We even came upon a floating island bolstered by dozens of plastic buoys used in oyster aquaculture that had solid areas you could walk on.[60]

Moore and his crew named the island HI-ZEX after the brand name of Japanese plastic fishing floats that comprised its bulk and put considerable effort into documenting its features with drone footage (figure 2.4), and even sketching a hand-drawn map . Moore, in a blog post from the time, was well aware of the irony: "For years I have been telling people that there is no such thing as a 'plastic island' in the Pacific Gyre. I now have a map of one that has aspects of permanence, a metal anchor 40 feet deep, solid rope beaches, some of which you can walk on as if you were on land."[61]

I can imagine marine scientists reacting to his words, eyes rolling, sighing to themselves, "Oh, not again. Why won't the solid island go away?" I can hear them telling me, "See, I *told* you Algalita was the source of all those

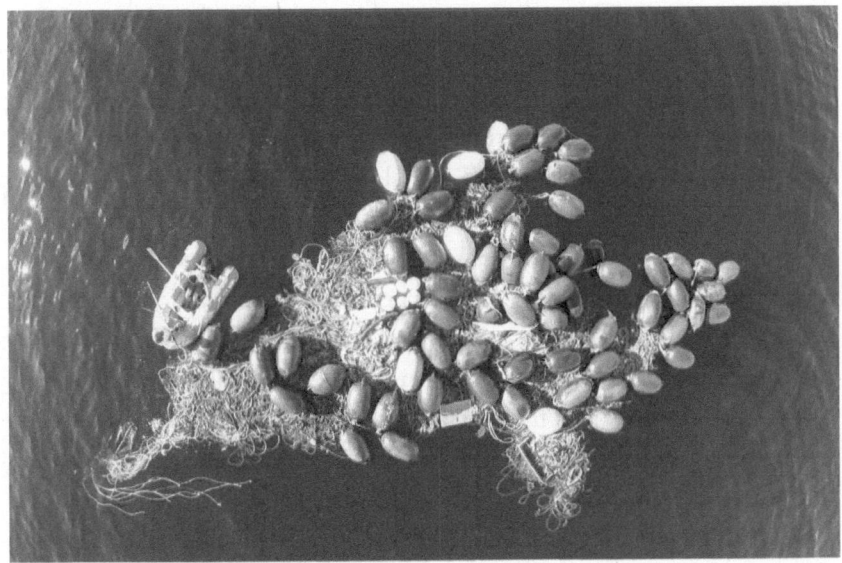

Figure 2.4

HI-ZEX Island, a conglomeration of namesake plastic aquaculture buoys, ropes, lost fishing nets, and all kinds of plastic. A zodiac is tied alongside the island's upper left. Still from drone footage courtesy of Algalita Marine Research and Education.

misconceptions." It is not only activists whose concerns are but built into ways of knowing that bring about the very things that come to matter. Technical critiques from the natural sciences seem to have generative potential as they are taken up as challenges to be overcome with optimism and ingenuity. And the allure of debunkery is not the domain of the natural sciences alone. As an STS scholar, poking fun at garbage patch development projects and other seemingly fantastical solutions is an easy way to gain the confidence of scientists or elicit laughter from audiences, especially when images of futuristic floating cityscapes are readily available. Ridicule proposals for developing robot zooplankton to clean plastic from the sea long enough, and the joke may be on you when, a decade later, you have to confront a world where JellyfishBot is swimming about before your book making a case against tech fixes for global plastic crisis sees the light.[62] Caring for missing things has a tendency to bring them into being.

The language of misconception, misinformation, and myth so prevalent among the scientific community and echoed by ChatGPT is often wielded to separate fables from facts, bad representations from good. Approaching trash island as a form of concern helps explain why it cannot be reshaped by adherence to scientific accuracy. Trash island has emerged inseparable from ocean plastic pollution as a global crisis. Form and concern, like matter and meaning, nature and culture, cannot be disentangled. Not with scientific accuracy, or better representations, or nets that scoop plastic from the sea. Where the island goes, so too goes an ocean plastic problem: They are fact-checked into shared obscurity; they are diluted into plastic soup only to congeal back into islands; they are built together as landforms with colonial aspirations. One way or another, ocean plastic pollution becomes meaningful (or not), a global crisis (or not), is addressed (or not) in concert with its island form. Attempts to reconcile representations with material reality seem to only widen other gaps, between science and action, knowledge and publics, or form and concern: They "remain trapped in binary oppositions."[63] In trash island's absence, fewer hurdles for the territorial expansions of fossil fuel industry's extractive worlds; in its presence, the ultimate *terra nullius*.

A synthetic frontier in the most literal sense—new ground made of plastic—trash island shows how the forms of things are entangled in power relations, not natural geographic features already there. Island forms, as Stefan Helmrich writes of waveforms, also "manifest as energies." They "substantiate anxieties, terrors, and sometimes, optimisms about the shape of history and of the future."[64] As an absence, trash island is a missing thing, a landmass that cannot be located, but it is also a process: It is made and unmade in the encounters that aim to disappear it, change it, create it. Trash island is at once history repeating itself and an anticipatory object "shaped by the not yet of the future."[65] The contingencies of emergence make space for alternatives to the seemingly inevitable place-making expansions of extractive worlds. It becomes possible to take a stand against oppression if you "cut differently the shape of a thing," reminds Puig de la Bellacasa.[66] Islands already have many other forms for those wishing to listen—ones that have been fluid, connected, relational, all along.

Strict adherence to modern measurement and material accuracy in isolation can also make concern, and even matter itself, disappear. The trash island is not the only thing that might be missing. Plastic relations become disentangled to the point where a meta-study arrives at the surprising conclusion that "ninety-nine percent of ocean plastic has gone missing."[67] A team led by Carlos Duarte, oceanographer at the University of Western Australia, surveyed global oceans in 2010 and 2011, collecting trawl samples much as we did on the North Pacific expedition, and combined their results with existing estimates including Goldstein's. Whereas models based on surface samples estimate the amount of plastic floating in global oceans in the tens of thousands of tons, those based on global plastic production and the probability of plastic ending up in the ocean estimate the amount to be in the millions of tons. Their conclusion is striking: The models fail to account for the vast majority of plastic expected to be floating in the ocean.[68] As for where the missing plastic has gone, there are theories that it has sunk out of view under the weight of plankton, taken to the skies in the bellies of birds, is circulating as nanoparticles in the clouds, and coursing in your own blood stream.

ENTANGLEMENTS

Intermittent encounters with plastic interrupt the steady rhythms of sailing, breaking the flow of life on the boat and the boat through the water. After so many days of just holding on, it is with much excitement that we finally begin to see plastic: a red crate, a black tube, a yellow float. Each sighting is meticulously negotiated (Plastic or jellyfish? Green or yellow? Twelve inches or eighteen?) and logged with GPS coordinates. While conditions have improved, it is still too rough to collect scientific samples. The manta trawl—a net designed to be towed alongside the boat—requires calms seas and slow movement; waves make it bounce across the surface, precluding the contents from becoming trusted representations of discrete slices of the sea. In the meantime, a device designed for higher speeds is skipping alongside the boat, gathering ocean plastic that can be used for education and outreach purposes. Marcus improvised the high-speed trawl from duct tape and spare parts when a previous expedition captain refused to slow down for science, but today it is the weather and waves that are not cooperating.

A sudden commotion rouses even the sleeping night watch team, bringing everyone on deck. We've spotted a netball: a tangle of lost fishing nets, ropes, line, and whatever else it gathers along the way, which in this instance appears to be us. The weighty plastic mass has snagged the high-speed trawl and is threatening to sink it. In this same moment, a fishing rod trailing hopefully behind the boat elicits the metallic zip of unspooling line. Plastic has caught us, and we have caught a fish. In a rush of activity accompanied by cranking winches and flapping sails, the crew scrambles to slow the boat, save the trawl, and land the fish. Some grab cameras and others grab knives, all as forward bouncing motion gives way to sideways rocking that tests our balance anew. The celebration of mahi for dinner (first dissected for science as the cameras roll) is temporarily diminished by the netball escaping, until it too is secured to the boat, now tethered alongside on our terms.

Our resident marine ecologist and film crew dive into the still-churning seas to investigate. They return with photographs of an eerily beautiful tangle, like an ocean plastic tumbleweed. Ropes and lines, white, yellow, orange, blue, green, and black, are all knotted together and adorned with the odd shred of plastic film. Even more surprisingly, the netball is full of life, a floating synthetic reef. Dozens of fish congregate in the shadows below where our dinner mahi had very likely been seeking a meal of its own. Sea slugs and even oysters are nestled amongst nylon strands embossed with algae. Their lives here are precarious. Though these creatures are understood to belong in the ocean, they would not otherwise be found so far out to sea. But the plastic is both shelter and snare. Netballs are also known as ghostnets for the tendency to keep fishing in the absence of humans (figure 2.5). Measuring

Figure 2.5
Fish both dead and alive with the tangled plastic of ghost nets beneath HI-ZEX island in 2014. Video still from "Mausoleum" courtesy of Algalita Marine Research and Education.

six feet in diameter, this one is too large to keep, but over the next few days, we bring aboard other floating objects thick with life. Gooseneck barnacles with heads buried deep in a hunk of yellow foam. Metallic blue spheres of fish eggs laid on a plastic bag. An oyster adrift on a nonslip bathmat coated in brown algae. Crabs that skitter across the deck, having abandoned their floats. And the blue crate of fish that forever changes how I think about responsibility. Plastic-saturated entanglements of life and death.

3 LIVING IN THE PLASTISPHERE

In the shadows of trash island lurks something far more tentacular: jellyfish with plastic embedded in their bodies; bacteria metabolizing synthetics for energy and nutrients; whole populations of ocean insects not merely surviving but flourishing with plastic. When I set out in pursuit of synthetics on the *Sea Dragon*, I was not at all expecting to encounter communities of marine life inhabiting, traveling, and even thriving with plastic. A desolate trash island loomed far closer on the horizon of possibility than any kind of living relations. Wasn't the entire expedition motivated by the harms plastic poses to oceans, not merely as an aesthetic afront to visions of pristine environments but as downright dangerous to lively bodies? My shipmate, marine ecologist Dr. Hank Carson, was not surprised. He had joined the expedition with the explicit intent of studying the emergent relationships between floating ocean plastic and marine life. Midway through the voyage, Carson was giving me a tour of the tiny, improvised laboratory space below deck on the racing sailboat turned research vessel: a sleeping berth with the mattress removed to form a workbench. Wedged beside a bread maker, with his laptop perched on a freezer that filled with scientific tissue samples as it was emptied of meat for human consumption, Carson was taking digital microscope images of invertebrates found on plastic objects the crew had plucked from the garbage patch earlier in the day. It was here, as I gawked at the lavender spiral of a sea snail magnified against a stark white background, that I first heard the term "plastisphere." The plastisphere, as Carson explained, is the name that marine ecologists have given to communities of organisms that live attached to, or associated with, plastics. "It's like the Anthropocene for marine biology," he continued, drawing a parallel with the geologic age that attributes to the human species epochal powers previously reserved for natural phenomena like meteor strikes and volcanic eruptions. Where the Anthropocene calls on global sedimentations of human products as stratigraphic markers for a new geological epoch, the plastisphere calls on human products to define a new realm of life as a site for biological inquiry.

The place-making expansions of synthetic frontiers are proliferating in more forms than massive islands or contiguous continents. With the plastisphere, synthetic fragments measured in millimeters are being distinguished from surrounding seawater as new kinds of territory for marine life and scientific study alike. "Plastic serves as a novel ecological habitat in the open ocean," announced the team of researchers credited with formally introducing the term plastisphere to the scientific literature in 2013.[1] The marine microbiologists had been cataloging organisms living on plastic fragments less than five millimeters across, the size of pencil erasers, not states. Their inhabitants are measured in microns, a unit commonly called on to describe the thickness of sheets of paper or single strands of hair, spanning the limits of bare-eye visibility to blood cells and bacteria. The plastisphere has since grown to encompass macro as well as micro life: biofilms of diverse microbes and colonies of coral-like bryozoans adrift on synthetic fragments, but also the parade of sea snails, oysters, and nudibranchs under Carson's microscope. And while these first studies focused on plastic communities in the garbage patches, scientific usage of the term is no longer restricted to specific areas or even to oceans.[2] The plastisphere can refer to lively communities associated with plastic in any kind of aqueous environment, with researchers most recently mapping its expansions to terrestrial environments as they search for a potential soil plastisphere.[3]

As explored by ecologists, even microplastics parallel the formation and discovery of new lands as empty spaces into which life can expand.[4] Floating plastic is especially of interest to marine ecologists because it can support and transport organisms in unpreceded ways. Synthetic polymers stand out for being a recent, human-instigated, chemically distinct, and exceptionally durable addition to aquatic environments. Not only does plastic have a tendency to stay afloat much longer than other materials like wood or even pumice, it also hosts communities that do not exist in the surrounding seawater. Large plastic objects and fragments can be home to species not usually found in the open ocean, or in concentrations and arrangements that depend on plastic's presence, leading some researchers to draw direct parallel between floating plastic pieces and "islands."[5]

Moreover, plastic provides all kinds of opportunities for extreme "rafting," where life forms take long-distance trips by "hitchhiking" on durable floating materials.[6] The biological term for organisms spreading to "unoccupied habitats" is literally colonization. For example, in Western narratives of the Hawaiian islands, accumulations of lava that first rise above the waterline are "hot and lifeless as the surface of the early earth," surfaces that are "barren" until "colonized" by algae, lichen, moss, and ferns.[7] The plastisphere is similarly positioned as a repetition of the earth's evolution, with an entire new lifeless world of geos that comes to be filled with animate bios, bacteria emerging from the sea to inhabit plastic as life once emerged from the ocean to inhabit dry land: plastic as "a hydrophobic surface that promotes microbial colonization."[8] Bacteria, diatoms, and their associates similarly come to "settle" plastic, sometimes within hours of synthetics entering the water.[9] These narratives stand in stark contrast to Kānaka Maoli genealogies, such as the Kumolipo chant where creation emerges from a coral polyp and land is not bare geos, but an elder sibling to be cared for.[10] Distinctly Western anthropocentric politics are entangled with the boundaries of biological description. If coming to inhabit empty land is celebrated, organisms that successfully colonize new-to-them spaces to the detriment of those already established are declared invasive species, while those that undermine human industrial interests, by taking up residence on a ships' hulls or competing for resources in aquaculture for example, are accused of "biofouling," dirtying with life.

At the same time microplastic fragments become new territory for the expansion of life in miniature, the plastisphere concept actively scales from micron measurements to an entire global realm. The plastisphere is "analogous to the biosphere" and "rivals that of the built environment in spanning multiple biomes on Earth," argue ecologists.[11] The plastisphere expands the very ranks of elemental layers Western scientists commonly use to delineate the earth's systems into geosphere, hydrosphere, biosphere, and atmosphere. This division brings discrete kinds of substances into being that roughly correspond to land, water, life, and air with differences preceding, thought certainly not precluding, their interactions.[12] Bios remains as ontologically distinct from geos as water is from land. Despite being a "new ecosystem" defined by associations, the term plastisphere predominantly describes the community of life *on* plastic: a "thin film of life" or "epi-plastic" coating, that mediates interactions *between* plastic and environment.[13] The assertion that plastic alone has formed a distinct geological layer in contention as a marker of an Anthropocene epoch has its own neologism: the Plasticene.[14]

The line between geos and bios, substrates considered empty until occupied by life, is powerfully entangled with Western ways of knowing and making extractive worlds. As articulated by Elizabeth Povinelli, the Western divide between bios and geos has long differentiated forms of existence that are or were once alive from the never alive. This separation of living and nonliving entities attributes some arrangements of matter the potential to be born, to reproduce, and to die, while rendering others inert.[15] For Povinelli, this dominant privileging of biological life over geological resource constitutes a gap that tears apart entangled beings. She refers to this "in-between" as the Carbon Imaginary, bringing emphasis to the deep entrenchment of vitalism in metabolic processes by the natural sciences and the humanities alike.[16] Derived from fossil fuels, the molecules forged into synthetic plastic were once ancient life, the majority of which was once alive as marine plankton. Yet, synthetic plastics predominantly remain carbon based only in the most atomic-elemental sense, as industrial processes that crack and reform chemical bonds seemingly remove plastic from organic life cycles as evidenced by their uncanny refusal to biodegrade on human timescales. Plastic, as Heather Davis argues, is matter constituted to "exists outside the cycles of life and death while nonetheless acting on them, embodying an existence akin to the geological."[17] Even in its plastisphere associations, plastic becomes an artificial surface to be inhabited or extracted.[18]

At stake with the line of bios/geos is no less than the contemporary constellation of power that Povinelli calls settler late liberalism: the arrangement of techniques for controlling markets in the face of social justice and anti-colonial challenges that emerged in the mid-twentieth century.[19] As many have argued, modern formations of power have long been dominated by biopolitics, a form of governing populations that works by making some bodies live while leaving others to die. What Povinelli astutely attends to is how biopolitical techniques require working agreements of what counts as bios in opposition to geos in the first place: "Western ontologies are covert biontologies—Western metaphysics as a measure of all forms of existence by the qualities of one form of existence (*zoe, bios*)—and that biopolitics depends on this metaphysic being kept firmly in place."[20] So too does colonial entitlement to land for extraction, settlement, or contamination require maintaining the living/nonliving divide that distinguishes those to be governed from those to be used.[21] Mountains are reduced to piles of rocks and minerals, rivers to ecosystem services, never living relatives, Earth Beings, ancestors, or kin.[22] Deprived of lively agencies, geos cannot be heard to speak for or defend itself. Moreover, denying Kānaka Maoli recognition of the lively authority of natural elemental forms is coupled with the insistence that recognition can only be sought from nation-states.[23] The living/nonliving divide is crucial to capitalist relations as "industrial capital depends on and, along with states, vigorously polices the separations between forms of existence so that certain kinds of existents can be subjected to different kinds of extractions."[24] Those who fail to see the world in dominant Western terms are not trusted with enacting the living/nonliving divide (or drawing the land/water line); so many Indigenous peoples have their ontologies reduced to quaint beliefs, cultures deemed living fossils to be managed accordingly.[25] At the same time, plastisphere relations, like Anthropocene conditions more broadly, shake these foundations in new ways; making visible the impossibly of maintaining life/nonlife separations.[26]

In this chapter, I approach the plastisphere as a synthetic frontier where the place-making lines dividing bios and geos, living and nonliving, continue to be challenged and remade. The plastisphere in particular furthers vital conceptualization of the expansive energies of synthetic frontiers, enfolding crisis into sustaining growth. What follows explores dominant narratives that establish plastic as antithetical to life and on whose terms. I show how in both the laboratory and a decades-long public cleanup campaign, forms of making and sharing knowledge about ocean plastic pollution are animated by an insistence on the separation of plastic and life. Strategies for change that continue to shore up divides by insisting plastic and bodies should not meet are being challenged anew with the emergence of plastivorous life forms. Capable of metabolizing synthetic plastics, not just taking up residence on their surfaces, plastivores are gnawing away at a material previously assumed to be impervious to life. In the process, they are being celebrated as the missing link in an industry-sanctioned vision of a circular plastic economy that renders plastisphere life itself extractible. This vision not only intentionally leaves petrocapitalist relations of violent expansion intact buts holds the potential to further naturalize them when wielded as evidence that the earth is evolving to take care of synthetic waste.

DEADLY ASSOCIATIONS

In my surprising first encounter with the plastisphere, I was by no means alone in assuming that plastic-life entanglements were undesirable, if not straight up deadly. Anti-plastic sentiment was and remains overwhelmingly energized by a sense that plastic is harmful to life, whether in the form of larger objects, microplastic particles, or unseen chemical flows.[27] Take entanglement, which for scientists, educators, and activists working with ocean plastic pollution, is negative by definition. As a technical marine biology term, entanglement refers to marine life ensnared by plastic debris: fishing line slowly strangling seals, six-pack yokes muzzling dolphins, and bottle cap rings giving hourglass figures to turtles. These are physical relationships—contact between bounded animal bodies and plastic materials—that in the words of marine biologist David Laist, "put animals at a survival disadvantage."[28] Entanglement causes "serious injury or death," as "many animals, if not most so caught, find it difficult to escape entanglement and are doomed to drown or die from injury, starvation and general debilitation."[29] With documented impacts on at least 243 marine species, ocean pollution entanglement is a serious threat to marine biodiversity.[30] Given such plastic's "lethal potential" with potentially extinction-causing consequences, it seems to follow that plastic and marine life *should* be kept separate. Indeed, many articles in the scientific literature use their conclusions to support ocean cleanup initiatives and associated technological solutions, "Marine clean-up initiatives are integral given the longevity of plastics in the environment," conclude a team of zoologists studying seals.[31] Dis-entanglement is possible and desirable.

The most viral popular images of ocean plastic harm also feature animal bodies in deadly, or at the very least painful, associations with plastic. Chris Jordan's now iconic photographs of dead Laysan albatross chicks on Midway Atoll in the North Pacific leave audiences gasping in shock and shame. Image after image of decomposing bird bodies reveal colorful piles of plastic fragments in what was once a belly, recognizable objects like lighters and bottle caps that could very well be *your* plastic, scooped from the sea and fed to baby chicks by devoted parents. Mistakenly eaten or ingested, but unable to be properly digested, plastic is not only outside of but impeding life processes.[32] With or without picking up the reference to senseless albatross murder moral allegory courtesy of "The Rime of the Ancient Mariner," the message distilled is that human hubris has brought death to innocent wildlife even two thousand miles from the nearest continent (Midway is so-named for being halfway between Asia and North America).[33] Then there's the now infamous sea turtle filmed as a plastic drinking straw was removed from its nostril. The video, which topped 110 million views of the original alone by April 2024, was made and shared by a Texas A&M graduate student researcher and her team who had been carrying out fieldwork on ridley sea turtles when they came upon an animal that seemed to be in distress. At first, the researchers assumed a barnacle or other sea creature had taken up residence in an unfortunate location. As the extraction proceeds, initial concern turns to shock and disgust with the realization that the culprit was a plastic drinking straw.[34] The turtle's suffering has been leveraged in support of legislation including plastic straw bans with appeals to viewers who empathetically cringed along with the blood-tinged pain of the straw being removed with pliers.

As concerns about concerns about microplastics and even smaller nanoplastics gain traction, anxiety about plastics crossing living body borders has taken increasingly human forms. The winning submissions from the 2023

Plastic Kills! short horror film contest, a collaboration between the USC Annenberg Hollywood, Health and Society program and the nonprofit Plastic Pollution Coalition, all focus on plastic in human bodies.[35] In a runner-up entry, a woman is mocked by her best friend for saying plastic water bottles are "literally killing us," only to be very literally murdered by a giant disposable water bottle. Making the message even more explicit, the short concludes with fear-inducing text that reads: "Disposable plastic water bottle waste literally kills 1.1 million marine creatures a year. Who will be next?" The catalogue of threats to wildlife in the ocean "out there" become threats to the "in here" of human bodies. The visceral horrors of plastic in human bodies permeates the audience choice winner, "No Escape from Within," where a woman eating alone on a dark evening, with a shadowy human-shaped specter hauntingly backlit in the window, swallows the broken tines of plastic fork after fork on repeat before vomiting them back up in a disgust-inducing display of flesh-shredding, bloody plastic. The overall winner "Blastic," features a vampire sucking down a plastic bag of blood while watching the news, where he learns of plastic contamination in human blood. The ensuing quest to purify his supply provides a campy, yet poignant, critique of dominant "solutions," from swapping out personal plastic items for other materials, to proper disposal and redoubling litter cleanup efforts. Alas, such actions are futile, plastic too nefarious, bodies too permeable, and plastic blood proves fatal even to vampires.

Plastic is not only harmful to individual bodies and species, or even to biodiversity, but appears to be killing the very planet itself. Visions of plastic as deadly to wildlife and humans all too easily escalate into visions of a plastic planet as a dead planet. The before images of Recycled Park in chapter 2 depict a river of plastic as devoid of life; dead albatross chicks become "casualties at the frontlines" of a "planetary emergency"; and headlines have announced plastic is killing marine microbes that produce a large portion of the earth's oxygen.[36,37] In popular conversations, this is commonly expressed as worries that plastic is "killing the planet" or even "destroying the earth," both of which Google readily autofills. "Planet or Plastic?" challenge headlines, with the implication they are fundamentally incompatible.[38] A plastic planet seems to almost always be a world without humans: plastic so durable it will outlast us all. The leap from the end of some ways of life, to the end of all life as the end of the world, echo dominant patterns of Anthropocene dystopias and the status quo–preserving logic of contemporary crisis.[39] As many scholars have noted, Western end-of-world narratives again and again express threats to some human ways of life as the end of humans a species, or even the end of life or the world entirely.[40] Povinelli intriguingly suggests that such extinction fears themselves are underpinned by the fear of bios returning to the inert "nothing" of geos from whence it once emerged. Bios is so privileged over geos in Western ontologies that the threat of extinction, of a planet without (some) humans, or without life, is equated with not having a planet at all.

SEPARATION SCIENCE

Like land/water, the categories of living/nonliving are physically inscribed on saturated worlds, including in the making of knowledge about them. Following the ocean plastic samples I helped collect on the *Sea Dragon* led me back to Hawai'i, completing another lap of the Pacific, this time by air.[41] I caught up with the samples at Hank Carson's laboratory space at the University of Hawai'i, Hilo. His postdoctoral research setup was nothing fancy, a bit of bench in a shared office and a back room for sample processing, but it seemed luxurious compared to the converted berth on the boat. Carson had an informal arrangement with Algalita, the nonprofit who had sponsored the expedition, where he would initially take the samples for his own research, thus saving the organization the difficult task of raising funds for the much less glamorous work of sample processing. As a marine ecologist, Carson's main plastic project was quantifying the diversity and density of organisms living on plastic fragments. In practice, this involved pulling out a random selection of ten fragments from each sample and preparing them for the scanning electron microscope—a process that rather ironically involved coating the synthetic fragments and the remains of their micro-inhabitants in gold. A student researcher gave me a guided tour of the resulting images, zooming in until I could see the remains of bacteria and diatoms (one image featured a sphere-shaped bacteria on an irregularly wrinkled background that I might have otherwise guessed was a human egg in a fallopian tube). The rest of the process was one I was now very familiar with: sorting the complete sample jars to separate plastic from plankton in order to calculate the density of plastic floating in specific areas of the ocean. Algalita's lab manager, Gwen Lattin, had herself been to Hilo to train the student researchers, one of whom I joined for the afternoon (which led to what is likely my only appearance in the acknowledgments of a marine ecology journal article).[42]

My own training in sorting ocean plastic samples took place at Algalita's California laboratory behind the SEA Lab, a hands-on ocean science center managed by the LA Conservation Corps. Its modest rooms housed Algalita's archive of plastic samples from oceans around the world and two dedicated staff members equipped with humble instruments and extraordinary patience. Nearly every surface was covered in all manner of plastic waste: Samples in process are gathered near the sink, bagged vials of finished samples are perched on top of a metal rack, and white bags of beach trash mysteriously occupy much of the available floor space. An informal display of specimens congregated near the window: jars of mixed plastic and plankton from different gyres, bags of plastic-laced "sand" from Kamilo Beach. Here, technicians and volunteers painstakingly processed surface samples of mixed-up plastic and organisms sifted from the open seas. With the aim of quantifying and monitoring the distribution of plastic in the ocean rather than studying the intricacies of life on plastic, Algalita researchers did not describe their work using the term "plastisphere." They did, however, calculate the quantity of plastic in relationship to the quantity of life. The plastic-to-plankton ratio is behind the widely circulated claim that plastic outnumbers plankton 6:1 in the North Pacific[43] and similarly contributes to the projection that plastic will outweigh fish in the ocean by 2050. In practice, this involves quite literally weighing one against the other. But first they must be separated as they circulate the space of the lab.

Working with forceps under dissecting microscopes, staff and volunteers physically sorted samples collected a sea into two glass jars. "Basically, it's plastic versus everything else," explained Algalita's lab manager. Though most commonly labeled "plastic" and "plankton" (figure 3.1), these containers had a telling variety of synonyms: Plastic was sometimes called manmade; plankton was interchangeable with animal, organic, natural, and even real.

Figure 3.1
Ocean surface samples collected aboard the *Sea Dragon*, sorted into plastic and plankton in Hank Carson's Hilo laboratory. Photograph by the author.

These categories for sorting matter not only contain assumptions about what belongs floating in the ocean (plankton/organic materials) or not (plastic/human products), they also map natural/artificial divides as separations between living/nonliving entities. That living entities in the samples are now dead, casualties of quite ordinary scientific knowledge production, is not the point; they were once alive and still exist within the realm of the biodegradable. The plastic-plankton ratio, continued the lab manager, was a comparison between nutritive and nonnutritive materials floating in the open ocean, or the likeliness a body looking for a meal would get something it could digest. Underpinning the categories for sorting samples was a living/nonliving divide that privileged the metabolic processes of biological life while assuming plastic was beyond them.

Plastic/plankton, artificial/natural, nonliving/living are not metaphorical boxes; they are containers for physical separation that ends with a material archive of dried plastic bits and dried plankton bits in separate plastic bags. Though I had read the lab protocol closely, the instruction to "remove all recognizable pieces of plastic" did not at all prepare me for the actual task of painstakingly separating plastic from plastic under the microscope. Though the lab manager had carefully selected a beginner's sample relatively less dense with tricky plankton, and I was tasked with looking for a common substance that ostensibly floats, picking out plastic was not so simple. My first look through the lens revealed a surprising world of strange sea monsters. The translucent slime that I assumed was pulverized jellies transformed into a crowd of microscopic creatures: symmetrical pieces with bulging eyes, segmented bodies, and hairy appendages, too endowed with recognizable qualities of lively bodies be plastic. To my relief, I also spotted a few brightly colored fragments recognizable as plastic, which I gracelessly chased around a Petri dish in an attempt to remove them with forceps. But plastic did not always come in bright whites, blues, and greens, or float cooperatively in plain view. With the lab manager carefully supervising and sharing the kind of advice that comes with years of experience, I learned to look where plastic hides in the meniscus at dish edges, how to stir the sample to unstick pieces from the bottom and from each other. With more practice, I learned to distinguish tatters of clear plastic film from the fingerprint-like markings of fish scales and fraying synthetic line from tapered zooplankton appendages.

Attempting to disentangle a microscopic ocean one teaspoon at a time, I confronted the fluidity of material cosaturations that refuse to conform to solid expectations and standard demands in frustratingly practical terms. I expected describing the samples to be straightforward: How many plastic pieces in the dish? What color are the pieces? But sun-brittle pieces shatter under the pressure of forceps, changing size and multiplying in number mid-process. Do I record the original number, closer to what was at sea, or the new number that is replicable if my work is checked? Wisps of clear film that threaten to take flight with the slightest breeze do not register as matter, even on the finely tuned scales. Fishing line tangles into pieces that are somehow one and not one at the same time. I worked out the difference between "white" and "transparent," then returned in the changed light of a cloudy day to find the distinction once again imperceptible. The seemingly innocuous sprinklings of potential plastic confetti that occupy very little physical space become hours-long projects of sorting and counting hundreds, and in extreme cases, thousands of fragments that must add up across data sheets. Categories and boundaries are enacted in this work of classification intended to script material flows, to determine future movements of sample bits and pieces from jars into Petri dishes, archives, and publishable figures.

Yet, as I experienced at both Algalita and Carson's labs, plastic's entanglements continued to exceed both practices and categories. As I worked, I overheard another volunteer describes the contents of her Petri dish as "a plant with legs on it"; moments later, another volunteer hesitated and conferred with her colleague before depositing a fragment (or is it a fish tooth?) in a jar labeled "plastic." Even those with years of experience and marine biology training were quick to provide examples of specimens troubling the sorting process: pieces of plastic film camouflaged by algae that cannot be scraped off; or salps, a jellylike filter feeder, with plastic on their insides (figure 3.2). Should the plastic be removed? Was it already part of them, or were plastic and bodies merged by the force of water through the net used for sample collection? In which jar do they belong? The boundaries between what counts as life and what does not are constantly breaking down and being remade, even as practices of *separating* plastic and life are enacted to assess how they might be *connected*. Sorting practices are constitutive of more than knowledge: They produce plastic and marine life as ontologically distinct forms of matter that belong, or not, in bodies aquatic, organic, or scholarly. The mundane acts of producing scientific knowledge about relationships enact place-making binary differences; the cuts of life from nonlife bringing into being the impact of plastic *on* the environment/ocean/marine life/biodiversity.

PLASTIC SPECIES

If tracing plastic's disentanglements in the laboratory shows how science based on separation attempts to brings worlds into being founded on plastic and life as discrete substances that should not meet, such foundational

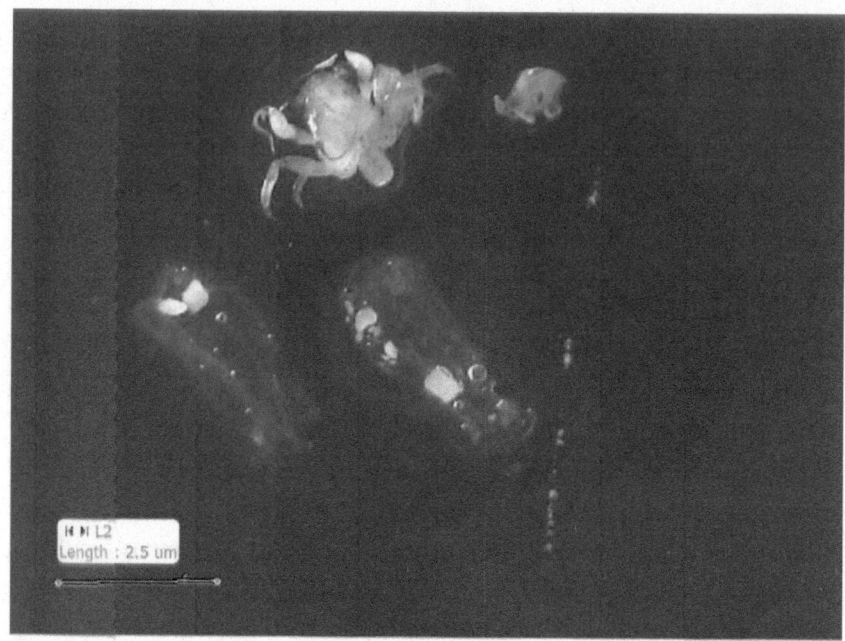

Figure 3.2
Two salps with plastic fragments in their bodies (and a tiny crab above). Courtesy of Algalita Marine Research and Education.

separations resonate broadly through major anti-plastic pollution campaigns and their imperatives to separate living and nonliving "species." Not only do bodies of plastic creatures move through the space of the lab, "novel associations of plastic and marine life" can also be found traveling in the poster and video campaigns of government and nonprofit organizations, aiming to educate the public about the problems of synthetic pollution. As I became more familiar with plastisphere communities proliferating at sea, I began to notice all kinds of proliferating plastic and life entanglements: birds with cigarette beaks, taxonomies of bottle fish, plastic bags described as agents with power to impact wildlife. Presented as "dangerous" and "non-native species," these images of plastic creatures were meant to encourage the cleanup of plastic from the sea, the very possibility of their category-crossing existence an argument for untangling plastic from life. Living/nonliving boundaries of species became implicated in the life and death politics of the work of conservation in the face of Anthropocene threats of mass extinction.

The walls of the Algalita lab were covered with an array of written and visual materials: plastic identification flow charts, reminders not to wash human dishes with the science sponge, and advertisements for pollution-related events. One slightly faded poster caught my attention each visit. Labeled the "Cig Egret," it featured a mean-looking heron on the beach, head seamlessly transitioning to tubular white and orange tobacco-product beak: The bird has a butt where its mouth should be (figure 3.3). Produced by the California Coastal Commission, the poster is an advertisement for the annual coastal cleanup day, part of a state-led public education program mandated to "engage the public in protection and restoration activities."[44] Despite appearing as paper, cigarette filters can splinter into thousands of microplastic filaments and contain toxins that do not biodegrade. By number of individual items, they are one of the most common objects found during beach cleanups.[45]

The California Coastal Commission is a state agency formed in 1972, to "to protect, conserve, restore, and enhance the environment of the California coastline."[46] For the most part, this involves regulating coastal land and water use, but the agency's activities also includes public outreach. Inspired by an event in Oregon called "Plague of Plastics," the agency began organizing an annual public cleanup day beginning in 1985. In the decades since, over 1.6 million participants have removed over twenty-six million pounds of trash from beaches and waterways in California alone. Taken up by the DC-based nonprofit advocacy group Ocean Conservancy, the event has been expanded to nearly every state and over one hundred countries globally. Topping one million participants in 2018 alone, the International Coastal Cleanup "has become the world's largest volunteer event related to the marine environment."[47] At the annual cleanup, trash is not only collected but sorted and counted by type—an accumulated long-term dataset that can be used to inform policy, business, and individual decisions to "reduce both plastic production and pollution." The resulting data, for example, was used in support of the California statewide plastic bag ban in 2017. There is growing understanding that plastic pollution is a systemic problem requiring

Figure 3.3
The Cig Egret "Non-Native Species of the California" poster advertising the California Coastal Commission Cleanup Day 2005.

solutions that exceed individual actions and recognition of watershed connections that expand what counts as a "coastal" cleanup far inland.[48]

The California Coastal Commission maintains an online archive of annual cleanup event posters dating back to 1986. Visually, the Cig Egret appears as the descendant of the 2003 campaign posters that featured stylized local (read: native) animal species, charismatically worthy of care and free of trash: a cautious raccoon, a skinny-legged avocet, a bright red newt, and a snappy crab. In 2005, the crab morphed into Spork Crab, the avocet into the Cig Egret, and the series as a whole into the "Non-Native Species of the California Coast." In contrast to the shock tactic photographs of deadly plastic relations, like the images of decaying Laysan albatross chicks, the Cig Egret and its synthetic kin play with the possibility of disturbing ways to be *alive* with plastic (and described by the CCC as "surreal").[49] Cig Egret acts as a monstrous animal other defining what counts as a legitimate life form through deviance from species as logical types and scientific categories (not to mention the political implications of being declared "non-native" by the state of California). Their power, however, depends on public recognition that such plastic-crossed creatures are unnatural, an argument positioning bodily entanglements of plastic and animal life as a problem for which cleanup is the appropriate solution. Framed by the worn edges of a well-thumbed trading card, the posters link the desire to collect with the imperative to clean up. To collect waste is to keep such illegitimate species from becoming reality. The creatures featured in the posters embody unnatural entanglements of life and nonlife, making categorical (im)possibilities publicly visible.

Over the following decade, the always visually striking posters trace an evolution from animal to plastic itself as species. Modified in subsequent years, suspect plastic creatures lose their trading card frames. The images become sharper in resolution, more photographic in presentation, and decidedly less imaginary in effect. The updated posters prominently display the numbers of the offending items in question collected at cleanups to date: 5,066,669 cigarette butts, 1,102,042 bottle caps, as if each item collected prevented another monstrous mutation. By 2010, when another poster series makes its debut, coastal creatures have become completely replaced by colorful photographs of artfully arranged plastic trash collected on Kehoe Beach in Northern California. "Help Reduce Trash" reads the text above a gathering of bottle caps and other materials separated by type. In the course of six years, the posters mark a transition from coastal animal species worthy of protection, through waste-animal entanglements, to sorted plastic "types" that have themselves become problematic. United—and at the same time divided—by the logic of "species" and the necessity of cleanup that provides continuity with each new series, kinds of coastal wildlife become kinds of trash.

This visual evolution of plastic species is explicitly realized in the 2013 CCC campaign. With playful irony, each poster in the series not only organizes types of trash but names them as genus and species, stamped with the rallying imperative: "Let's make trash extinct" (figure 3.4). Cigarette butts, now "animal" in their own right, not merely disturbing appendages, are united as an entire taxonomic family of "Sandlocked fumers," whose members include "*Smokus discardus*," "*Jointus acutus*," and stereotypically gendered lipstick-stained "*Ladies leftebehindus*." In bestowing Latin(ish) binomials on trash, lively qualities appear to be extended to plastic. Well versed in new traditions of awarding agency to any entity capable of making a difference, at first I read such posters as doing just that, drawing a parallel between animal and plastic capacities as a productive way to remember the unintended consequences of synthetics, that they too have lives and are active participants in the shaping of seascapes.[50] These posters, and many like them, however, draw their power from the ethico-political impossibility of just that: Plastic is not and should not be equated with species.[51] Scare quotes are enlisted to emphasize the ironic use of a way of naming and categorizing restricted to entities endowed with the qualities of biological life. This campaign relies on and produces audiences that know that kinds of plastic waste do not count as kinds of life. If living species inhabiting plastic were a surprise, plastic species are an ontological impossibility.

Simultaneously, the campaign very seriously draws on recognition of human Anthropocene agency to render life forms extinct. An expansive bestiary of twenty-six illegitimate trash species endures on the CCC website, with each entry providing succinct justification not for "species" protection

Figure 3.4
Cigarettes as plastic species in the "Let's Make Trash Extinct" poster advertising the California Coastal Commission Cleanup Day 2013.

but its eradication. These are species that *should* go extinct, and the campaign directive describes "how you can make it happen!" It is humans, and not plastic, that have the power to shape environments, make landscapes, and control the destiny of *all* life, an ultimate restoration of faith in human power over plastic and the very possibility of containment. In calling for plastic's "extinction," the campaign playfully deploys biopolitical control to reinforce a foundational life/nonlife divide to secure plastic in its proper place: safely quantified in datasheets, contained in trash bags, and evicted from natural environments. Plastic is not inert in the sense that it sits in place doing nothing. There is recognition of its ability to travel without humans from upstream watersheds to the sea and recognition of chemically active toxic flows. Plastics "escape into the environment," where "Eating them could cause an animal to choke or starve to death because the plastic isn't digested." These kinds of activities, however, are not to be confused with metabolic life processes that they are understood to threaten. The boundaries distinguishing life and nonlife are broken down only to be reconstituted in ways meant to motivate particular forms of responsibility. As with the scientific studies, the public images at once connect and separate organic bodies and plastic ones. The tensions between kinds of plastic and kinds of life become explicit arguments that position separation—once again cleaning up synthetic materials from the environment—as the dominant course of action.

PLASTISPHERE LIFE

From the Latin ranks of their preliminary taxonomy of life on plastic in the garbage patch, the authors of the first scientific article to name the plastisphere in print called attention to a particular grouping of microorganisms they found "especially intriguing" because they were "pitting the surface" of a sample.[52] While there were many types of life on microplastics, the roundish bodies of these bacteria in particular were a very close match for the shapes of the divots they were occupying on the synthetic materials. The accompanying grayscale scanning electron microscope image shows spherical cells, clustered like pomegranate seeds in pith (though, zoomed out a bit in another image, the arrangement appears more like an unfortunate rash to the uninitiated). Determining whether the bacteria were actively making the indentations or merely inhabiting them based on a fortuitous perfect fit was beyond the scope of the study, but the authors noted that the cells appeared to be actively growing and reproducing on the plastic.

The plastisphere's lively associations challenge assumptions that subtend plastic's status as nonliving and deadly matter, tripping up the easy leap from plastic presence to bodily harm or planetary death as a return to geos alone. With relationships that are supportive of life, not just in harmful contact with it, emergent plastic-life relations wedge open ontological cracks in the nonbiodegradable. Fish bites fray edges, bacteria pockmark impervious surfaces, barnacles burrow into fissures, fungi chew through industrial-strength chemical bonds. In a basement laboratory at the Scripps Institute of Oceanography, researchers investigate relationships between plastic presence and *Halobates*, a marine water skeeter that is the only open ocean insect. With eggs that need to be anchored to a floating platform, their reproduction is limited by the availability of once much rarer hard substrates on the ocean surface. After plucking insect eggs one by one from microplastic samples under dissecting microscopes, the results point to a very strong statistical correlation with the amount of plastic at sea: *Halobates* have the audacity to not only survive but downright flourish with ocean plastic at the level of population.[53]

This is by no means an apology for plastic. Plastisphere flourishing does not counterbalance the unequally distributed violences of colonial expansion, petrochemical extraction, and enduring chemical toxicity. Rather, plastisphere life demands careful consideration of how plastic harms are mapped and on whose terms. Even the widespread knowledge of Laysan albatross chick deaths by plastic arguably further distances responsibilities. As Jordan himself recounted to the BBC, the images generated a massive outpouring of emails from people wanting to go to Midway Island to save the chicks as a "trauma response," a rush to save animals in paradise instead of focusing on where the plastic is actually coming from (much like the response to footage of Kamilo Beach after the Japan tsunami).[54] Moreover, as Max Liboiron points out, dominant narratives about albatross victims exceed scientific knowledge: The bird deaths may very well have been natural, and Laysan albatross populations are generally stable as a species.[55] For Liboiron, the widespread use of Jordan's images focuses on the wrong relations: exploiting albatross suffering to further narratives of plastic bodies meaning death, instead of recognizing albatross as figures of resilience in worlds where chemicals and bodies cannot be kept separate.[56]

The expansiveness of frontiers stems from not only maintaining but creatively reorganizing boundaries to ensure ever-more entities align with the logics of endless growth. As Povinelli argues, the foundations of bios/geos so long largely unquestioned are becoming increasingly visible and increasingly shaken by Anthropocene challenges. Attributing the human species, a category of living beings, geologic capacities messes with longstanding arrangements that reserved agency for (some) humans who acted upon but were not themselves geos. Humans have gained new kinds of agency, and geos new kinds of potentiality, but critical theory is not always rising to the challenge.[57] Even feminist new materialist projects that make great efforts to extend agency to all kinds of entities (my own work included) arguably leave intact the very grounds upon which agency is anchored: vital assumptions about what counts as life itself. It is not enough to question plastic's shifting relationships to nature or to award plastic lively agency. As Davis observes, "Animacy is not itself an anathema to waste colonialism," as oil endlessly turned object fuels petrocaptialism.[58] Indeed, Povinelli provocatively suggests that on this front capitalism may itself be the most radical: "Nothing is inert, everything is vital from point of view of capitalization."[59] Ocean plastic arguably constitutes the first planetary-scale crisis for the plastics industry, one with a billion-dollar response to addressing it.[60] But plastic that can be metabolized subverts the boundary threats to endless accumulation, securing future petrocapitalist expansions of synthetic frontiers.

NATURALIZING SYNTHETIC CIRCULARITY

Confirmation that bacteria was, as a scientific matter of fact, metabolizing plastic, emerged not from the sea but from the primordial ooze of a waste facility in Japan. A team of researchers lead by Kohei Oda, a microbiologist at the Kyoto Institute of Technology, had been collecting plastic-contaminated soil and sludge in pursuit of a microorganism that could help soften the surface of synthetic fabrics all the way back in 2001.[61] What they found was evidence of previously unknown bacteria that were not only physically breaking apart the long, repeating molecular chains of plastic polymers like polyester but fully metabolizing the component parts: The microorganisms were using synthetic plastic as a source of energy and nutrients necessary for fueling the processes of life. It was more than a decade before publication of the term plastisphere, even before "trash island" first appeared in popular news coverage of the garbage patch, and due to a lack of concern for plastic, the results were not published at the time.[62] Fifteen years later, with plastic accumulations endowed with the status of a global crisis (and after a few other research teams had come to similar conclusions about a few exceptional bacteria and fungi),[63] Oda and his colleagues published their ongoing plastic-eating bacteria research, based on analyses of samples collected at a Japanese plastic-bottle-processing facility, in *Science*. Through laboratory isolation, the team of researchers confirmed that the bacteria, *Ideonella sakaiensis,* were using polyethylene terephthalate (PET for short, one of the most common consumer plastics and the material of ubiquitous disposable water bottles) as a "major carbon source for growth."[64] At least one type of synthetic plastic was not only a novel substrate in the biological sense of a surface to inhabit but in the biochemical sense of a substance an enzyme (a type of protein that speeds up bodily processes, including digestion, in all living things) can reform into new products. The bacteria were metabolically dependent on plastic.

Where the scientific discovery of plastivores was from the start caught up with industrial interests, their entry into global conversations is from the start caught up with the aspirations of a plastic circular economy. The article in *Science* does not stop at describing new kinds of plastic-life relations. Rather, it quickly transitions from the study of "consortiums" of materials and organisms to isolating specific enzymes bacteria produce as "enzymatic machinery" as a "platform for biological recycling of PET waste products."[65] In subsequent scientific articles and popular news coverage of them that follow, coupling plastivores with recycling is the norm, not the exception.

Media coverage was quick to speculate about bacteria and insect larvae's potential for "taking a bite out of the recycling problem" (Cornwall 2021), offering a "viable solution to the plastic problem" (Mulhern 2021), or even "saving the planet" (Ap 2016). With an understanding that plastic pollution is a waste management problem for which recycling is the unquestioned solution, plastic-degrading microbes are articulated again and again as a biotechnical fix for accumulations of plastic waste. Provided they possess the capacity to metabolize synthetics, plastic-eaters were no longer harbingers of the end times—they were a remedy for planetary plastic woes.

With plastivores caught up in dreams of endless synthetic circularity, plastic-eating becomes something to be celebrated. The party is not for the creative persistence of life despite petrocapitalist accumulations but in service of its continued expansions. As Oda later told a reporter from the *Guardian* chronicling ongoing efforts to solve the plastic crisis with the help of microbes, "I say to people, watch this part of nature very carefully. It often has very good ideas," (2023). As the reporter continues, "Many scientists have come around to Oda's view—that for the host of seemingly intractable problems we are working on, microbes may have already begun to find a solution. All we need to do is look." And many researchers are looking. A recent large-scale study mining (their verb) genomic datasets identified a potential thirty thousand existing plastic-degrading enzymes, an abundance that increases with the amount of plastic pollution present in a given space. The authors laud these microorganisms' "great potential to revolutionize the management of global plastic waste," while simultaneously holding up their results as evidence that "the earth's microbiome may already be adapting."[66]

As asserted in a feature article of the Association of the American Society for Biochemistry and Microbiology, nature provides not only resources for endless plastic circulation but "offers a blueprint."[67] Though evolution appears to be so very conveniently aligning with petrocapitalist human plans, this natural process must be efficiently harnessed for sustainable management. Microbial "carbon workers" are deemed to work too slowly without technoscientific intervention.[68] It is only by extracting the enzymes and bioengineering them to degrade synthetics more rapidly that plastivores are shown to be fully capable of "closing the loop of the circular economy."[69] The potential success of plastivore recyclers requires the separation of enzymes from microbial communities to redirect plastic into a seamless loop from which materials supposedly cannot escape. "The nice thing about enzymes is you get the building blocks back," explains one structural biologist. "That's potentially an infinite process, so it's really attractive."[70] Where plastic waste extracted from the sea by massive nets gets molded into the foundations of new land, plastivore enzymes promise the recovery of molecular elements. Plastivores are being enlisted to subvert not only the immediate threat concerns about ocean plastic pollution pose for endless petrochemical expansion but overcome the very end of oil as a nonrenewable resource.

The industry-sanctioned vision of a circular plastics economy is no more a closed loop than islands are natural isolates. Recycling becomes yet another form where environmental conservation furthers colonialism. Tracing the trajectory of circularity in EU plastic policy debates, Sandra Eckert asserts that revolutionary potential has become so "coopted by industry to fit narrow vision of recycling as circularity" that the model is "really closer to linear."[71] Dreams of "closing the loop" have become reduced to eliminating landfills and ramping up recycling but without limiting extraction, toxicity, colonialism, or growth. For sociologist Alice Mah, "The circular economy for plastics is both a corporate battleground for containing environmental crises and a catalyst for intensifying expansion." A plastic circular economy is just the latest in a long string of "corporate strategies of containment and proliferation represent attempts to 'future-proof' capitalism against existential threats to public legitimacy, masking the implications for environmental justice."[72] Industry dreams of circularity further elemental separations and extractions meant to uncouple plastic's material persistence from entangled violences, in part, by focusing on technical problems and promises at the expense of alternate visions. Plastivores become a biotechnical fix for further entrenching the status quo, another appeal to sustainability that helps contain threats to petrocapitalism to secure the continued expansion of synthetic frontiers.

WHAT PERSISTS

Anti-plastic strategies that depend on imperatives to uphold nonliving/living, plastic/species, natural/culture divides are especially vulnerable to the expansive adaptability of petrocapitalism. With biotech visions of synthetic waste efficiently excreted back into circuits of production, organic and plastic life cycles merge without irony. If plastic can be biodegraded, if other species can productively incorporate it into bodily processes, then maybe it isn't so deadly after all. Moreover, being metabolized potentially undermines plastic's prodigal endurance, as its uncanny capacity to last seemingly forever is enfolded back into more familiar life cycles of decay and regeneration. If the material can now be properly biodegraded, why worry about producing and consuming less of it? Is there even a plastic problem at all?

While plastivores may seem like a promising antidote to the persistence of plastic as matter thought to be immune to biodegradation, plastic problems are not simply a stubborn solid material quality in isolation, and taking care of them cannot be limited to the realm of waste management alone. Following the lead of queer theorists helps disrupt appeals to nature as some kind of organic whole that can be enlisted to perpetuate the status quo, while simultaneously calling on humans to assume responsibility for the coevolution of bacteria and plastics production.[73] In this case, the status quo entails the continuation of petrocapitalist production and circulation of toxins; it frames plastic problems as a matter of waste management for which tech advances taking advantage of the "natural" evolution of recycling is the fix. Bioengineering enzymes from plastivores is not a break from these relations but arguably a means to further naturalize them in the face of a growing anti-plastics movement. Plastivores are a solution only when the problem is narrowly framed as a stubborn physical quality of synthetic materials in isolation; they do not address broader relations of domination and control that constitute some kinds of entities as extractable and manipulable in the service of (some) humans.[74] There is no biotech fix for the violence of petrocapitalism because it is at the same time a social and political problem with deep ontological foundations. Plastisphere flourishing does not in itself dismantle oil pipelines, undo intergenerational chemical harm, return land

and sovereignty to Indigenous peoples, or bring into question the most basic terms of associated conversations.

Plastivores can help instigate a radical shift in understandings of how to live with plastic, but this involves actively resisting the allure of petro-captialist material circularity to challenge the endurance of interconnected social and political relations. Taking care of plastic means attending to all its relations, oil and plastivores, for Zoe Todd and Heather Davis, respectively, as "paradoxical kin" or even "toxic progeny" with whom reciprocal responsibilities, not extractive cuts, must be assumed.[75] In staying with the trouble of what persists, enduring harms of chemical additives—conveniently named persistent organic pollutants—provide an appropriately microscopic fault line with the potential to fracture biotech visions into possibilities for something else. Plastivores cannot currently biodegrade all the many kinds of plastic, and especially not the lingering toxic chemical additives associated with them, and might never be able to do so. Endocrine-disrupting additives alter biological reproductive systems, interrupting the reproduction of broader social and political norms to which they are bound. In this way, "microorganisms provoke alternative conceptions of what material transformations involve . . . as an articulation of new collectives brought into the space of material politics."[76] Plastivores are neither simply a natural development releasing humans from obligations nor another natural resource to exploit. The task is caring for emergent relations without exacerbating associated harms.

LANDINGS

The promised calm seas finally arrive, the elusive high-pressure system that shifts the center of the gyre. We collect samples, marveling at the rare recognizable items captured by the small mouth of the trawls towed alongside the boat: bottle caps, a toothbrush, and most unlikely of all, a gray plastic toy gorilla figure. On one day when the winds had dwindled until the sea surface is barely rippled, shimmering steel gray with overcast skies, we are allowed to swim. The crew is giddy with the novelty of getting off the boat even for a few minutes, but most of all with the prospect of swimming in the middle of the Pacific Ocean. I struggle to find words for the color blue I see diving from the deck into water that imperceptibly fades into the depths. The closest land above the surface is still Hawaiʻi, now 1,200 miles away; the seafloor is some three miles below.

I do not become one with the ocean. While I manage to stay on the boat, aside from the one very intentional swim, life at sea continues to challenge my stomach's resolve. I type notes and blog posts lying face down, eyes closed, on the galley bench. Some fare better. Others fare worse, accepting with quiet reluctance, their fate of vomiting daily for the duration. The blue crate, fishy inhabitants evicted for good, is now home to a small collection of salvaged floats, foam, and bottles, all stripped of barnacles to minimize the smell. It is nearly three weeks on, and fresh provisions and fuel supplies are both dwindling. There are no more leeks in the lab cooler. The narrow passageways that once required navigating boxes of papayas and potatoes stowed on the floor are now clear. I still remember the day a crew member was caught surreptitiously eating the last tomato without sharing; the day we ate the last eggs. Everyone is getting restless, even loopy. A suspect report of "sea bats" mysteriously appears in the official logbook overnight. My watch team, now nicknamed the Wolf Pack, has taken to howling at the moon. One morning, I hallucinate a mirage of mountains as the sunrises over clouds layer dark and light that looks more convincingly like land than when we actually see it.

The science stops at the invisible edge of the US EEZ (exclusive economic zone), two hundred nautical miles from shore, as we do not have requisite permits to continue sampling. Various projects completed as best we can, given the conditions, we change course toward Canada. A major readjustment of the sails completely reverses the tilt of the boat, resulting in a symmetry of bruises and vastly rearranged space below deck. The bunk where I had nestled in relative comfort safe against the hull now lists about precariously, threatening to eject my sleeping body onto the floor three bodies below.

Fueled by teamwork, wind, and more diesel than anyone aboard would like to admit, there is once again land in sight, orcas between the islands in the straight. Then a government dock in Victoria, British Columbia, a tidy provincial capital city adorned with flowers and the kitsch of empire. In the late dark of a summer evening, we wait on the boat for customs officers to arrive and check our passports upon our arrival, not to Canadian waters but to so-called Canadian land. Though the last fresh produce was weeks ago, I can hardly eat the vibrant fresh salad once ashore, the rocking sensation of the sea so strong I am now landsick (a term my spellcheck refuses to recognize). A week later, laying down and closing my eyes will set my body rocking with the memory of waves. Years later, camping under a metal and canvas canopy in a strong breeze whisks me from the Southern California desert back to sea; I awake in the night ready to spring on deck and adjust the sails. I have not become one with the ocean, but my embodied view from land is no longer unmarked, even if it takes many more years before I can express it in these terms.

Our last leg on the Sea Dragon is through the Strait of Georgia, still named for the King of England, part of the Salish Sea, only formally recognized as such by Canadian settlers since 2009. The water is brown with silt carried by the Fraser River, the shores green with the familiar forests of my childhood. Amid barges laden with logs and gravel, and container ships whose contents we can only speculate, we turn the *Sea Dragon* in a slow, full circle. The autopilot, it turns out, was not exactly broken after all. It simply needed to be recalibrated—a process completed in ten minutes in anticipation of heading onward from Vancouver with a skeleton crew. The first mate sheepishly admits that keeping it out of service had the benefit of keeping us all occupied, especially through the long night watches. Was that wild night learning to sail nothing more than a carefully curated drama ensuring guest crew had An Experience?

There's a press conference at the aquarium, the day after we arrive in Vancouver. A representative from Coca-Cola is among those in attendance, keeping one step ahead of us activists. We murmur to each other with knowing looks. We have good reason to be suspicious of industry intentions. Major players in the disposable plastic bag industry were at that very moment suing ChicoBag, a California-based maker of now commonplace reusable shopping bags. ChicoBag's founder was being taken to court for false advertising by industry officially taking issue with the matters of fact being used to promote the reusable product, among them, the claim that the garbage patch was a "landfill" twice the size of Texas (though industry attention was far more likely because of ChicoBag's activist roots and the impending California-wide plastic bag ban).[1] My watchmate Marcus has been, among his many other duties at sea, compiling scientific evidence to counter these accusations in court by establishing that plastic pollution is in fact a legitimate crisis for the ocean, and not just one for the plastics industry.

I summon the courage to speak with a reporter from the *Vancouver Province*, one of the major local daily newspapers, hoping my words will not come back to haunt me. She is very kind and seems to have done her research; she is, to paraphrase, aware of the media misrepresentation of trash island. Reassured, I do my best to describe my experience, emphasizing how larger pieces of plastic were very dispersed, how when you're out there it mostly appears to be clean blue ocean, until you look really closely and put nets in the water. There's no scary trash monster or trash island. What's really frightening, I continue, is the presence of tiny bits in sample after sample, day after day, no matter how far we were from land.

The expedition makes the front page of the newspaper the next day, along with a photo of our densest sample held out as evidence by Marcus. The jar of bits in the picture—from a sample where if you look closely among the rainbow fragments you can see the gorilla, a pen cap, a little crab—most certainly contains plastic (the same jar on display in figure 2.2). But it contains so little of the experience of looking for a garbage patch in the middle of the sea; it falls so short of having meaningful conversations about it. About what it feels like to be so far from land that the sighting of a single bird brings everyone on deck while ocean plastic sightings have long become routine. About the cold, hard work it takes to haul the trawl back onto the deck of the rocking boat in the middle of a wet night watch to check for fish. Or about the difficult decisions those onboard face when recounting the voyage once back on land: Feed the current story and hope it sparks change, or struggle to explain exactly why you paid so much money to acquire a sprinkling of well-traveled trash confetti. Then I see it. The thick, bold letters marching across the front page read: "'Frightening' Amount of Plastic Floating in North Pacific Gyre."[2] Collective embodied experiences, my carefully chosen words, become the stuff of alarming headlines.

SYNTHETICS: PLACING TOGETHER/WITH

Did you see it? The trash island? They ask, eyes wide with anticipation having learned that I have sailed through the North Pacific Ocean. Been there. Seen the infamous Great Pacific Garbage Patch with my own eyes. I still hesitate before answering, not to bask with the satisfaction of someone about to deliver The Truth, but because my answer nearly always disappoints: I don't know. But isn't it twice the size of Texas, they counter. I have seen photographs, they insist. I used to respond with a more confident negative; however, caring for missing things tends to bring them into being. Matterings, including island-shaped presences and absences, are never complete. I am not offering a synthesis of humanities and natural sciences knowledge in some epistemic equation that settles the form of ocean plastic pollution once and for all; I am arguing for accountability with the synthetics of making and knowing that are always relations to place.

The lines of this story insist on unfolding in circles. Straight paths of origin and progress bend with currents re-turning and re-turning in a widening gyre. The trash island appears and disappears, again and again. Uncountable plastic migrations around, through, and with all kinds of bodies escape the carefully labeled containers of science and the measured knowledge they are intended to produce. Plastic transcends the limits of nature, only to become it again as plastivores are recruited to further naturalize petrocapitalist evolution. My words merge into the headlines of my own media analysis.[1] The *Sea Dragon* loops, recalibrating embodied experiences. Three weeks/ thirteen years pass between getting on the boat and landing. Re-turning,

Barad writes, refusing to contain the violent physics of atomic Pacific histories in the past, is a radical "circling back around," not as means of reflecting but "reconfiguring what might yet have been."[2] Radically reconfiguring synthetics is to refuse the resolution of tensions between descriptions and reality into objectively better representations. Re-turning unsettles a geometry of strict boundaries and fixed points, a geography where plastic endures without transformation.[3]

Re-turning also allows for a reckoning with earthbound geos as the unmarked elemental standard for measuring and writing relations with place (geometry, geography, even geontologies).[4] Plastic's tendencies toward paradoxical flows are at odds with terracentric expectations: enduring while falling apart, embodying a solidity of ongoing chemical transformations, wreaking harm while instigating novel ecologies, all while caught up in nonlinear process of circulation, saturation, and disappearance. Still, plastic is far too often approached as solid waste, as stable, bounded objects that can be kept or taken back from "the environment."[5] Dominant solutions continue to reinscribe the line between land and water with all kinds of barriers. The Ocean Cleanup's most recent concession that plastic pollution must also be addressed at the source does not look like regulating fossil fuel extraction, decreasing plastic production, or decolonization; it looks like "interceptors": floating booms and conveyor belt systems that extract plastic where rivers meet the sea.[6] Even my six-year-old daughter's ocean plastic picture book maps "a border of awareness" in a crusade to "stop the plastic flow" along North America's Pacific coast, protecting the ocean from land. The most well-meaning environmental projects can keep producing the line that plastic (or rising sea levels) is not supposed to infiltrate.

Plastic's persistence exceeds a physical mismatch of solutions, just as the trash island exceeds misrepresentation. As ongoing enactments of deeply embedded Western ontologies that expand synthetic frontiers, the garbage patch cannot be remedied with more "correct" science or journalism. Plastic demands accountability for the elemental terms upon which solidity and fluidity are understood, where islands come to matter. An accountability that continues to question: With whose topographies are more just worlds being mapped and constituted? Indigenous peoples are not trusted with distinguishing where land ends and where water begins, with recognizing living from nonliving entities, or determining what is fixed and what moves.[7] Synthetic frontiers are ongoing violences against other ways of islands and other ways of water.

Black, Indigenous, and Pacific Islander scholars are already leading conversations offshore in a refusal of the cramped spaces of bounded isolates. In *The Black Shoals*, Tiffany Lethabo King refuses the default analytic terrain of continental philosophy through her conceptualization of the shoal. Shape-shifting with flows, only temporarily above the surface if at all, a shoal is a formation that confounds the boundaries and binaries of the line: It "is simultaneously land and sea," she explains.[8] While the term settler colonialism took momentum from Hawaiian activist and scholar Huanani-Kay Trask's nuanced grappling with genocide, King argues that the concept has been usurped by "white" variants that sideline conquest and genocide by recentering the settler.[9] Swirling sedimentations become spaces of radical Black and Indigenous solidarity that demand "Man give an account of his violence."[10] To shoal is to insist that abolition and decolonization are incommensurate with the unifying dreams of reconciliation through diversity and inclusion.

A trash island, like other reclamation projects that take plastic back from the sea, materializes a form of reconciliation that leaves intact the terraforming synthetics of a colonizing world. Accountability for plastic's relations to place entails a refusal to take as given the most foundational terms of ocean, plastic, pollution, and crisis. Even before Epeli Hauʻofa's rallying call for an Oceania united by a sea of islands, poet Vernice Wineera was countering extractive configurations of an empty ocean. As read by Alice Te Punga Sommerville, Wineera enlivens the edge of the ocean with the presence of elemental connection, rather than reducing it to the beginning of the void or the end of history.[11] Continuing to refuse the cutting edges of the line, Vicente Diaz, Filipino-Pohnpeian scholar and writer from Guam, has declared "No island is an island."[12] He pulls Indigenous epistemologies and cartographies beyond the literal shoals of the reef "to destabilize the stubborn definitions of *land* that remains unmarked."[13] Here, fluid seascapes are questionings of the very elemental lines themselves with "a region and tradition of traveling that trouble hard lines between land and sea as well as between many other categories typically understood to be diametrically opposed to one another."[14] It is not enough to substitute water for land; the task requires reconfiguring their ontological grounds "to combat exclusive categories of self and other and the bounded territoriality on which they are affixed."[15] Other ways of islands already exist; they are already being lived.[16] They are already catalyzing new chemistries, new trajectories, and new forms of abundance.[17] Places made together/with seams of relationality.

And so I return to a place where this book might yet have begun. Not searching for a spectacular garbage patch continent in the middle of an ocean wilderness. Not marveling at the colorful trash islands plotted in science fiction. But with a very ordinary beach walk in Long Beach, California, where synthetic frontiers have been looming in plain sight all along. It is a scene that I biked past on my regular commute to the Algalita office: the sandy stretch of shore known as Alamitos Beach. Separated from the Ports of Los Angeles and Long Beach by a marina, and by the line of the famously channelized Los Angeles River meeting the sea, local knowledge of associated contaminations means that only tourists tend to swim here.[18] It's golden hour on a clear fall day, and I'm walking along picking up bits of trash, but especially looking for plastic stories. The light is just so, illuminating objects on the beach, the spray of waves crashing, and ships and islands in the distance alike. I snap a photograph juxtaposing plastic pollution widely seen as disposable consumer goods—a single-use white plastic fork—with that more subtly present in other forms: a small chunk of fragmenting Styrofoam, and a sprinkling of nurdles, the preproduction pellets that are industrial waste (figure 4.1). There are two nurdles visible in the photograph just above the fork handle, but many more congregate in the tideline. Some are bright and smooth, others worn rough by sun and waves, and tinted yellow by the chemical accumulations of synthetic toxins that stick to their surfaces.

What stands out to me now is the island hovering on the horizon. It is one of the four THUMS islands, named for Texaco, Humble, Unocal, Mobil, and Shell (Humble is now Exxon), built with granite and dredged sand to access an undersea oil field within city limits. They are drilling platforms disguised to be more palatable in proximity to such a densely populated

Figure 4.1
Plastic along the shoreline in Long Beach, California, with a THUMS island and container ship on the horizon. Photograph by the author.

shore.[19] Designed in the 1960s by the theme park architects responsible for Disney's Tomorrowland, the islands are decorated with palm trees and artificial waterfalls, derricks hidden in white condo-like structures with blue balconies, all artfully illuminated at night.[20] Islands of aesthetic containment that permit petrocapitalist expansions. Beautification was explicitly part of the agreement with the city allowing the project to go forward. In a heroic intersection of the mid-twentieth-century frontiers of oceans and outer space, each individual island is named for a fallen NASA astronaut. As the sun sets over the oil rigs and port cranes, the container ships heading to and from the Pacific, I pluck a green plastic army man figurine from the sand, adding militarism to the entanglements of extraction, consumption, and waste continuing to shape the Pacific. Though at the time I vaguely grasped the complex relations, I did not have the words for synthetic frontiers. I did not know how to see the line even as I was biking along it, flying over it, sailing right past it. I was disappointed by the blurry fork in my image, when the camera had been bringing the tideline into focus, an ocean of connections all along.

I am still only beginning to glimpse what it means to see and to not see islands. To see and to not see trash islands as synthetic frontiers: terracentric expansions of extractive worlds now in the name of sustainability. To question not only the promise of cleanup, but why its designs involve new coastlines in the open ocean, floating parks in rivers, or bioengineered microorganisms. To see how colonial entitlement to land logically extends to making it from what in the Western worldview is the middle of the largest watery place on the planet. To ask why the more-than-human inhabitants, even in miniature, are gathered into a whole global sphere open to new extractions that buoy up the old ones. Plastisphere life is either rendered extractible as a natural solution, or rendered expendable, killable in the spatial transgressions of becoming invasive.[21] Synthetic frontiers map all kinds of expansions where territorial lines of control that emerge with the products of modern science and technology enfold crises into yet more progress. Situating these techniques of power that render fluidities extractable, enclosable, and even inhabitable helps draw lines of connection between modern science and technology and the furthering of petrocapitalist landforms.

Synthetic frontiers do not have to be made of plastic. Place-making expansions of petrocapitalism have long been in embedded in the Port of Long Beach, as detailed by Christina Dunbar Hester, even before the THUMS islands raised the seafloor into a dry platform for drilling. San Pedro Bay was shaped already in the 1930s by the imposition of the line of land/water on a "useless wetland" transformed by global flows of fossil fuels.[22] The expansions of a harbor by and for oil animated by the ever-mutating destruction—the violent ecologies—of petrochemical infrastructures. Coastlines awash in plastic are being challenged anew by rising sea levels that undermine the very line of territory. Seawalls themselves become infrastructures of petrocapitalism (even more so when they are molded from recycled plastic).[23] Even without plastic, islands can continue to be made and to be made frontiers of science and technology. The myth of the isolate is alive on Guam, where bioengineered mosquitos are released on local populations and corporate "community engagement" means focus groups for anticipating and containing opposition.[24] Places themselves become experimental subjects, Seung Hee Cho argues, such as Jeju island in Korea becoming a testbed of permanent transformation by wind power.[25] With the emerging figure of the energy island, "landfills of the future" mean new land built from waste deposited at sea forms the foundations of renewable energy.[26] Energy humanities and petrocultures researchers map synthetic frontiers with the expansions of petrochemical infrastructures that impose the (pipe)line and resistances to them.[27] Terraforming extends underwater, as Ocean, Blue, and Hydrohumanities scholars are already doing the work of mapping ongoing expansions below the surface.[28] Violent separations extend to the sea floor as deep sea mineral extraction is justified by the expansive needs of the green energy transition meeting global climate goals.[29] Synthetic frontiers are everywhere in critiques of the forms of environmental "progress" that remain anchored in modernist control of nature and the preservation of the status quo. But most of all, synthetic frontiers are the thread that weaves all of the above together into the illusion of progress: where a trash island is giving way to greener, smarter futures, and when energy transitions and biotechnological progress have become entangled with more plastic and more violence, not less.[30] The trash island that does not exist makes visible the place-constituting makings and knowings of all kinds of boundary relations, the violent cuts that are ongoing choices, including the anchoring of sovereignty to dry land.

In Long Beach the veneer of modernity seems to have run thin. The city is slated to halt oil extraction by 2035, a decade ahead of the targeted date for doing so statewide. The future of the THUMS islands is an ongoing conversation, with pitches for wildlife sanctuary, hotels, windfarms, and carbon capture technologies being floated about. Regardless of the outcome, water must continue to be injected into the oil wells to keep the city from sinking.[31] Green energy transitions, however, too often take the form of petrochemical expansions elsewhere. In Corpus Christi on the Gulf Coast, highly contested plans for five marine desalination plants seem to be going ahead, the first in the state of Texas. The facilities are officially a response to drought-exacerbated water shortages, with existing fresh water supplies tapped out. But in a city already home to the world's largest plastic production facility, it is no secret that the supply crisis is both of and for the continued expansions of petrocapitalism.[32] Through desalination, seawater is made to subtend place-making growth, with Exxon among those investing in the new infrastructures of fossil-fueled plastic futures.

Another flight to Hawaiʻi, this one in 2023. As the plane banks toward Oʻahu, I look down again at deep blue ocean brightening to cyan near the shore, at green hillsides fading into the mist of a sea that has taken to the skies. This time I see it. Not the trash island, but the line cutting together/apart where once I could have only arrived at land, naturally isolated in an unimaginably vast sea. I return to the harbor where I first boarded the *Sea Dragon*, taking photographs of the colonial land/water divide, trying to imagine an ocean that connects together/with, rather than separates. Maybe if I squint just right. Maybe if I keep listening. In Pearl Harbor, a glossy sheen of oil continues to seep from the USS *Arizona*, a spill too patriotic to clean up; in the hills above, jet fuel leaks from military storage facilities into drinking water.[33] Traveling back home to the Dallas area now on a red-eye, another line drawn in jet fuel, I awaken to a passenger excitedly exclaiming without irony, "We're back on land!" The pull of continental forms supersedes oceanic fluidities even while airborne among the clouds at thirty thousand feet. What shape is cut by the line that connects the underground sprawl of fracking wells reaching under my North Texas home to the ongoing contaminations of Oʻahu's aquifers? I continue to grapple with entangled elemental responsibilities. I continue to listen. Again, I look for an ocean that connects, rather than separates; for worlds of exploration without conquering; for islands that refuse to be contained.

Notes

INTRODUCTION

1. Thomas M. Kostigen, "The World's Largest Dump: The Great Pacific Garbage Patch," *Discover Magazine*, July 9, 2008; Justin Berton, "Continent-Size Toxic Stew of Plastic Trash Fouling Swath of Pacific Ocean," *SFGATE*, October 18, 2007; ABC News, "Hidden, a 3.5 Million Trash Heap Lies in the Ocean," *ABC News*, August 6, 2008; The Ocean Cleanup, "The Great Pacific Garbage Patch," https://theoceancleanup.com/great-pacific-garbage-patch/.

2. Chapter 2 recounts the stories of several associated headlines, including one in the Science section of *The New York Times* that proclaimed: "Afloat in the Ocean, Expanding Islands of Trash." The truism (or is it?) that the Great Pacific Garbage Patch is "twice the size of Texas" continues to circulate widely, including, as of October 2024, on The Ocean Cleanup "Great Pacific Garbage Patch" overview page. Even the less traceable claim, that the garbage patch is visible from space, is common knowledge enough that a plethora of articles have been dedicated to explaining that this is not, in fact, the case. For example, on NOAA's web page, "Debunking Myths About Garbage Patches;" Lindsey Hoshaw, "Afloat in the Ocean, Expanding Islands of Trash," *The New York Times*, November 9, 2009; National Oceanic and Atmospheric Administration, "Debunking Myths About Garbage Patches," April 11, 2024, https://response.restoration.noaa.gov/about/media/debunking-myths-about-garbage-patches.html; The Ocean Cleanup, "The Great Pacific Garbage Patch."

3. While plastic materials and disposability converge in new ways in post–WWII America, they are not inherently connected. The first synthetic plastics were substitutes for luxury materials (ivory, tortoiseshell), and the first disposables were made of paper (shirt collars). Jeffrey L. Meikle, *American Plastic: A Cultural History* (New Brunswick, NJ: Rutgers University Press, 1995), 64; Susan Strasser, *Waste and Want: A Social History of Trash* (New York: Metropolitan Books, 1999).

4. With gratitude to Miri Powell for reintroducing me to the map of Synthetica through her historical narration of plastic and empire in the Pacific. Miri Powell, "Synthetics of Empire: Plastics, Toxicity and Waste in the Pacific," Society for the Social Study of Science, 2023. For Meikle's detailed analysis of the map of Synthetica, see *American Plastic*, 64–67.

5. For an exemplary account of the necessity of resisting the conflation of colonialism with capitalism (despite their many affinities), particularly in justice-oriented antipollution projects, see Max Liboiron, *Pollution Is Colonialism* (Durham, NC: Duke University Press, 2021), 13–16.

6. Meikle, *American Plastic,* 73.

7. Meikle, *American Plastic*, 64–67.

8. Roland Barthes, "Plastic," in *Mythologies* (New York: Hill and Wang, 1972), 97–99. Barthes's *Mythologies* is equally well known among ocean scholars for infamously deeming the ocean "a non-signifying field" that "bears no message."

9. Heather Davis, *Plastic Matter* (Durham, NC: Duke University Press, 2022), 12.

10. Amanda Boetzkes, *Plastic Capitalism: Contemporary Art and the Drive to Waste* (Cambridge, MA: The MIT Press, 2019).

11. Liboiron, *Pollution Is Colonialism.*

12. Zakiyyah Iman Jackson, *Becoming Human: Matter and Meaning in an Antiblack World* (New York: New York University Press, 2020).

13. Sophia Roosth, *Synthetic: How Life Got Made* (University of Chicago Press, 2017).

14. Roosth carefully details the sinuous lineages of the term "synthetic" in modern philosophy and science. Among its often-contradictory manifestations, she notes that "synthetic" has at some point meant both inductive and deductive processes. See especially Interlude 5 "What Comes Before," for a succinct overview of synthetic philosophy. Roosth, *Synthetic*, 175–176.

15. *Oxford English Dictionary*, s.v. "synthetic (*adj.* & *n.*)," June 2024, https://doi.org/10.1093/OED/8737220608.

16. Gustav E. Mueller, "The Hegel Legend of Thesia-Antithesis-Synthesis," *Journal of the History of Ideas* 19, no. 1/4 (1958): 411–414.

17. Brian Burkhart, *Indigenizing Philosophy Through the Land: A Trickster Methodology for Decolonizing Environmental Ethics and Indigenous Futures,* (East Lansing, MI: Michigan State University Press, 2019), 17–20.

18. Burkhart, *Indigenizing Philosophy Through the Land*, 21–22.

19. Roosth beautifully traces the lineages from the synthetic chemical industry promise of "better living with chemistry" to projects of" making life better" with contemporary synthetic biology that expand the very limits of what counts as life. Roosth, *Synthetic.*

20. Except as philosophers of chemistry Bensaude-Vincent and Stengers clarify, it was actually yellow and in France, but that's not how the story usually gets told. Bernadette Bensaude-Vincent and Isabelle Stengers, *A History of Chemistry* (Cambridge, MA: Harvard University Press, 1996), 183–184.

21. Esther Leslie, *Synthetic Worlds: Nature, Art and the Chemical Industry* (London: Reaktion Books, 2005), 7.

22. Leslie, *Synthetic Worlds*, 10.

23. Here I am building on the work of many scholars connecting modern power to land relations, including Chandra Mukerji's articulation of modern state formation through practices of territorial power; Eve Tuck and K. Wayne Yang's articulation of settler colonialism as land relations; and Max Liboiron's account of pollution as land relations. Mukerji, *Territorial Ambitions and the Gardens of Versailles* (Cambridge, UK: Cambridge University Press, 1997); Chandra Mukerji, "The Territorial State as a Figured World of Power: Strategics, Logistics, and Impersonal Rule." *Sociological Theory* 28 no. 4 (2010): 402–424; Eve Tuck and K. Wayne Yang, "Decolonization Is Not a Metaphor," *Decolonization: Indigeneity, Education & Society* 1, no. 1 (2012): 1–40; Liboiron, *Pollution Is Colonialism*.

24. Joe Harris and Martín Morazzo, *Great Pacific Volume 1: Trashed!* (Image Comics, 2013a); Joe Harris and Martín Morazzo, *Great Pacific Volume 2: Nation Building* (Image Comics, 2013b); Joe Harris and Martín Morazzo, *Great Pacific Volume 3: Big Game Hunters* (Image Comics, 2015).

25. The garbage-patch-as-trash-island figures in a growing number of climate fiction (or "cli-fi") and magical realist works of fiction. In Nick Hayes *The Rime of the Modern Mariner*, (Western) human hubris is made manifest as a garbage patch trash mountain solid enough for ships to run aground. In Catherynne M. Valente's *The Past Is Red*, a terraformed garbage patch is home to the remains of humanity adrift on a climate-flooded planet. In Wu Ming-Yi's *Man with the Compound Eyes*, a tsunami sets a massive garbage patch island on a collision course with the Taiwanese coast. Nick Hayes, *The Rime of the Modern Mariner* (Viking, 2012); Ming-Yi Wu, *The Man with the Compound Eyes*, trans. Darryl Sterk (Vintage Books, 2013); Catherynne M. Valente, *The Past Is Red* (Tor.com, 2021).

26. Charles Moore and Cassandra Phillips, *Plastic Ocean: How a Sea Captain's Chance Discovery Launched a Determined Quest to Save the Oceans,* (New York: Avery, 2011). Charles Moore, interview with the author, 2012.

27. Robert H. Day and David G. Shaw, "Patterns in the Abundance of Pelagic Plastic and Tar in the North Pacific Ocean, 1976–1985," *Marine Pollution Bulletin* 18, no. 6 (1987): 311–316. Jim Ingraham, interview with the author, 2012.

28. For a comprehensive overview of Western scientific knowledge about ocean plastic pollution, see Anthony L. Andrady, *Plastics and the Ocean: Origin, Characterization, Fate, and Impacts* (Wiley-Blackwell, 2022).

29. There are thousands of types of plastic, far exceeding the seven recycling categories (based on most common use) molded on the bottoms of bottles and other packaging. Samantha MacBride, *Recycling Reconsidered: The Present Failure and Future Promise of Environmental Action in the United States* (Cambridge, MA: The MIT Press, 2011), xii.

30. Jan Zrimec et al., "Plastic-Degrading Potential Across the Global Microbiome Correlates with Recent Pollution Trends," *mBio* 12, no. 5 (2021); G. Lear et al., "Plastics and the Microbiome: Impacts and Solutions," *Environmental Microbiome* 16, no. 2 (2021)

31. Ocean studies scholars have recently been calling attention to aqueous spaces in all their dimensions not simply as surfaces. For particularly influential examples see: Stefan Helmreich, "An Anthropologist Underwater: Immersive Soundscapes, Submarine Cyborgs, and Transductive Ethnography," *American Ethnologist* 34, no. 4 (2007): 621–641; Melody Jue, *Wild Blue Media: Thinking Through Seawater* (Durham, NC: Duke University Press, 2020b); Philip Steinberg and Kimberley Peters, "Wet Ontologies, Fluid Spaces: Giving Depth to Volume Through Oceanic Thinking," *Environment and Planning D: Society & Space* 33, no. 2 (2015): 247–264.

32. Anthony L. Andrady, "Microplastics in the Marine Environment," *Marine Pollution Bulletin* 62, no. 8 (2011): 1596–1605.

33. Chemical additives included during production alter material properties, making a diversity of plastics that are, for example, harder, more flexible or fire-resistant. These additives do not always stay attached. The off-gassing of potentially toxic additives called phthalates (which are also endocrine disruptors) is responsible for new car smell and the aroma given off by new shower curtains, tablecloths, and other consumer goods. Susan Freinkel, *Plastic: A Toxic Love Story* (Melbourne: Text Publishing, 2011).

34. As recounted to by Algalita's then president, Bill Francis, and corroborated in person. Francis poured over the titles for the over six hundred papers to be presented at the conference and found that close to 70 percent made arguments directly addressing plastic. As he explained in frustration, "The fact is, something like 70% of marine debris is in fact plastic, and to ignore especially in light of UNEP calling it an emerging issue, to me just seemed ludicrous." Bill Francis, interview with the author, 2012.

35. Daniella Russo, "Plastic Pollution, Not Marine Debris!" Plastic Pollution Coalition, 2011a; Daniella Russo, "Wrapping up 5IMDC, Honolulu, Hawaii," Plastic Pollution Coalition, 2011b.

36. On the necessity of further unsettling deeply embedded Western ontologies, including the marine debris/plastic pollution duo, that have come to dominate so many ocean-crossing conversations particularly those surrounding the March 11 Japan tsunami see: Kim De Wolff, "'Floating Things' and Methodological Drift: Accounting for Haunted Materialities in the North Pacific Ocean," *Social Studies of Science* 54, no. 4 (2024): 536–556.

37. As Stephanie LeMenager argues in *Living Oil*, oil has become so inextricable from American identities that it constitutes a dominant petroculture that the majority are unwilling to change. Instead of imagining how to live without oil, visions are limited to imagining how to live more productively with it. Stephanie LeMenager, *Living Oil: Petroleum Culture in the American Century* (Oxford, UK: Oxford University Press, 2014).

38. Stephen Buranyi, "The Plastic Backlash: What's Behind Our Sudden Rage—and Will It Make a Difference?" *The Guardian*, November 13, 2018. Break Free from Plastic "About BFFP," Break Free from Plastic. Accessed February 24, 2023, https://www.breakfreefromplastic.org/about/.

39. Hiroko Tabuchi, "How a New Treaty Could Clean Up Plastic Waste," *The New York Times*, March 2, 2022.

40. As measured by the increase in journal article publications. For details, see Tania Rabesan-dratana, "Report Traces Surge in Ocean Plastic Studies," *Science* 372, no. 6548 (2021): 1249.

41. Jenna R. Jambeck et al., "Plastic Waste Inputs from Land into the Ocean," *Science* 347, no. 6223 (2015): 768–771; Diane M. Sicotte, "From Cheap Ethane to a Plastic Planet: Regulating an Industrial Global Production Network," *Energy Research & Social Science* 66, (2020): 101479; Beth Gardiner, "The Plastics Pipeline: A Surge of New Production Is on the Way." *Yale Environment 360*, December 19, 2019; Roland Geyer, Jenna R. Jambeck, and Kara Lavender Law. "Production, Use, and Fate of All Plastics Ever Made." *Science Advances* 3, no. 7 (2017): e1700782.

42. Fredric Bauer and Tobias Dan Nielsen, "Oil Companies Are Ploughing Money into Fossil-Fuelled Plastics Production at a Record Rate—New Research," *The Conversation*, November 2, 2021; Zheng Jiajia and Suh Sangwon, "Strategies to Reduce the Global Carbon Footprint of Plastics," *Nature Climate Change* 9, no. 5 (2019): 374–378.

43. Bauer and Nielsen, "Oil Companies Are Ploughing Money into Fossil-Fuelled Plastics Production at a Record Rate—New Research"; Zheng Jiajia and Suh Sangwon, "Strategies to Reduce the Global Carbon Footprint of Plastics," 374–378.

44. Jennifer Gabrys et al., eds., *Accumulation: The Material Politics of Plastics* (London: Routledge, 2013); Sharon Lerner, "Waste Only: How the Plastics Industry Is Fighting to Keep Polluting the World," *The Intercept*, 2019; Alice Mah, *Plastic Unlimited: How Corporations Are Fueling the Ecological Crisis and What We Can Do About It* (New York: Polity Press, 2022).

45. Dominic Boyer, *Energopolitics: Wind and Power in the Anthropocene* (Durham: Duke University Press, 2019); Cymene Howe, *Ecologics: Wind and Power in the Anthropocene* (Durham: Duke University Press, 2019); Hamza Hamouchene and Katie Sandwell, eds. *Dismantling Green Colonialism: Energy and Climate Justice in the Arab Region* (London: Pluto Press, 2023).

46. Joseph Masco, "The Crisis in Crisis," *Current Anthropology* 58, no. S15 (2017): S65–76.

47. Masco, "The Crisis in Crisis," S73.

48. Kim De Wolff, "Plastivores and the Persistence of Synthetic Futures," in *Living in the Plastic Age: Perspectives from Humanities, Social Sciences and Natural Sciences*, ed. by Johanna Kramm and Carolin Völker (Frankfurt: Campus, 2023); Trisia Farrelly, Sy Taffel, and Ian Shaw, eds., *Plastic Legacies: Pollution, Persistence, and Politics* (Edmonton: Athabasca University Press, 2021); Liboiron, *Pollution Is Colonialism*.

49. United Nations Environment Programme, "Historic Day in the Campaign to Beat Plastic Pollution: Nations Commit to Develop a Legally Binding Agreement," March 2, 2022, Available from https://www.unep.org/news-and-stories/press-release/historic-day-campaign-beat -plastic-pollution-nations-commit-develop; Tabuchi, "How a New Treaty Could Clean Up Plastic Waste."

50. Davis, *Plastic Matter*, 47.

51. Liboiron, *Pollution Is Colonialism*, 9.

52. Matt K. Matsuda, *"AHR Forum*: The Pacific," *American Historical Review* 111, (2006): 758–777.

53. Jacob Darwin Hamblin, *Poison in the Well: Radioactive Waste in the Oceans at the Dawn of the Nuclear Age* (Piscataway: Rutgers University Press, 2008).

54. Liboiron, *Pollution Is Colonialism*, 114.

55. Liboiron, *Pollution Is Colonialism*, 15.

56. Rabesandratana, "Report Traces Surge in Ocean Plastic Studies," 1249.

57. Liboiron, *Pollution Is Colonialism*, points specifically to the Jambec et al. study I have cited above, for its well-meaning but ultimately colonial focus. This is not at all unique to ocean plastic pollution. The very first formal scientific article naming the Anthropocene, for example, ends with a call for yet more scientific research and modern management (Paul J. Crutzen and Eugene F. Stoermer, "The 'Anthropocene,'" *Global Change Newsletter* 41, [2000], 17–18).

58. Patricia L. Corcoran, Charles J. Moore, and Kelly Jazvac, "An Anthropogenic Marker Horizon in the Future Rock Record," *GSA Today* (2014): 4–8.

59. New materialism, most broadly construed, is an umbrella term for a return to materiality in the humanities and social sciences after decades of analysis dominated by language, discourse, and values. Part of a broader ontological turn, new materialists generally share a conviction that meanings, ideals, culture, and society do not exist independent from physical stuff and embodied practices (Diana H. Coole and Samantha Frost, eds., *New Materialisms: Ontology, Agency, and Politics* [Durham: Duke University Press], 2010). Worlds creatively emerge from generative practices, instead of being preconstituted by bounded objects that divide neatly into modern dualisms of matter/meaning, mind/body, subject/object, and so on. Many new materialists, especially following Jane Bennett, *Vibrant Matter: A Political Ecology of Things* (Durham, NC: Duke University Press, 2010), extend agency to nonhumans to emphasize their vital, constitutive role in ethicopolitical worlds. In this book, I build more specifically from feminist and queer new materialist scholars, such as Karen Barad and Mel Chen, who persistently keep attention on the active production of difference in the makings of agency and worlds. Mel Y. Chen, *Animacies: Biopolitics, Racial Mattering, and Queer Affect* (Durham, NC: Duke University Press, 2012); Bennett, *Vibrant Matter*; Coole and Frost, *New Materialisms*.

60. Karen Barad, *Meeting the Universe Halfway: Quantum Physics and the Entanglement of Matter and Meaning* (Durham, NC: Duke University Press, 2007), 26.

61. Barad, *Meeting the Universe Halfway*; Donna Haraway, "Situated Knowledges: The Science Question in Feminism and the Privilege of Partial Perspective," *Feminist Studies* 14, no. 3 (1988): 575–599.

62. Jue, *Wild Blue Media*.

63. Barad, *Meeting the Universe Halfway*, 148.

64. For the necessity of focusing on cuts and exclusions in Barad's work, see: Gregory Hollin et al., "(Dis)entangling Barad: Materialisms and Ethics," *Social Studies of Science* 47, no. 6 (2017): 918–941. For Barad's situating of her own work among feminist, and especially Black and Chicana feminist, conceptualizations of difference see (Karen Barad, "Diffracting Diffraction: Cutting Together-Apart," *Parallax* 20, no. 3 (2014): 168–187).

65. Stacy Alaimo, *Bodily Natures: Science, Environment, and the Material Self,* (Bloomington, IN: Indiana University Press, 2010), xi; Alexis Shotwell, *Against Purity: Living Ethically in Compromised Times* (Minneapolis, MN; London: University of Minnesota Press, 2016); Davis, *Plastic Matter.*

66. Davis, *Plastic Matter*, 6.

67. Shotwell, *Against Purity*, 3.

68. Bruno Latour, *We Have Never Been Modern* (New York: Harvester Wheatsheaf, 1993).

69. Shotwell, *Against Purity*, 204.

70. Jue, *Wild Blue Media.*

71. Melody Jue and Rafico Ruiz, eds. *Saturation: An Elemental Politics* (Durham, NC: Duke University Press, 2021), 6–7.

72. Building from Gay Hawkins's excellent work on the ethics and politics of especially plastic waste, I trace ocean plastic's "contingent materialities," as multiple, emergent, but also situated. Gay Hawkins, "The Politics of Bottled Water," *Journal of Cultural Economy* 2, no. 1–2 (2009): 183–195; Gay Hawkins, *The Ethics of Waste: How We Relate to Rubbish* (Lanham, MD: Rowman & Littlefield, 2006).

73. Gabrys et al., eds., *Accumulation*, 5.

74. Bruno Latour, *Reassembling the Social: An Introduction to Actor-Network-Theory* (Oxford: Oxford University Press, 2005).

75. At the same time, this project was conceptualized at a moment where scholars tracing the circulating "lives" or "biographies" of things, were giving new attention to object afterlives and practices of discarding, now under the umbrella of Discard Studies (Max Liboiron and Josh Lepawsky, *Discard Studies: Wasting, Systems, and Power* [Cambridge, MA: The MIT Press, 2022]). Consumption and use were no longer accepted as end points of linear commodity chains. But in following plastic afterlives at sea, it soon became apparent that I could not be limited to commodity forms, or even the assumption of bounded objects.

76. Stefan Helmreich, *Alien Ocean: Anthropological Voyages in Microbial Seas* (Berkeley, CA: University of California Press, 2009); Stefan Helmreich, *A Book of Waves,* (Durham, NC: Duke University Press, 2023); Stefan Helmreich, "Nature/Culture/Seawater," *American Anthropologist* 113, no. 1 (2011): 132–144.

77. Eva Frederick, "Ninety-Nine Percent of Ocean Plastic Has Gone Missing," *Science*, January 3, 2020.

CHAPTER 1

1. Liz Barney and Michelle Broder Van Dyke, "Welcome to Hawaii's 'Plastic Beach.' One of the World's Dirtiest Places," *The Guardian*, January 10, 2020.

2. Barney and Van Dyke, "Welcome to Hawaii's 'Plastic Beach.'"; "Kamilo," *Wehewehe Wikiwiki Hawaiin Language Dictionary*, https://hilo.hawaii.edu/wehe/?q=kamilo#w2w2-57658.

3. The dominant framing of pollution as a litter problem is exemplified by the Keep American Beautiful organization and campaigns which focus on postconsumer trash carelessly tossed into the environment and further individualize responsibility for proper discarding.

4. This unidentifiability is no accident, but part of the designed untraceability that Heather Davis calls synthetic universality. Davis, *Plastic Matter*.

5. The organization, its technology, and founder Boyan Slat have racked up an impressive collection of prestigious awards, including United Nations Champion of the Earth in 2015; Forbes 30 Under 30 in 2016; Wired 25 people who will shape the next 25 years in 2018; and the 2021 Ocean Hero Award. The Ocean Cleanup, 2023. https://theoceancleanup.com/.

6. Max Liboiron, "How the Ocean Cleanup Array Fundamentally Misunderstands Marine Plastics and Causes Harm," Discard Studies, 2015; Sy Taffel, "Communicative Capitalism, Technological Solutionism, and the Ocean Cleanup," in *Plastic Legacies: Pollution, Persistence, Politics*, ed. by Trisia Farrelly, Sy Taffel, and Ian Shaw (Athabasca University Press, 2022).

7. "Oceans • The Ocean Cleanup," *The Ocean Cleanup*, https://theoceancleanup.com/oceans/.

8. Taffel, "Communicative Capitalism, Technological Solutionism, and the Ocean Cleanup."

9. "Oceans • The Ocean Cleanup."

10. The Ocean Cleanup, *2022 Annual Report*, 2022.

11. Boyan Slat, "How the Oceans Can Clean Themselves," YouTube, October 24, 2012, https://www.youtube.com/watch?v=ROW9F-c0kIQ&t=137s.

12. A proliferation of aqueous concepts are already in circulation, variously gathering social science and humanities projects around water and oceans in their many forms. Here, I am especially building from those committed to rethinking with (sea)water the very conceptual foundations of power, rather than taking water as a setting for human dramas (i.e., at or on the sea). For this particular project, I am using the term hydrohumanities to emphasize the ontological and epistemological inseparability of water and power, where "human-water-power relations are irreducible to their component parts" (Kim De Wolff et al., *Hydrohumanities: Water Discourse and Environmental Futures* [University of California Press, 2022], 6). At the same time, I use the term hydrohumanities to attend to the productive confluences of what are too often salty/fresh divergences separating river and ocean scholars into separate streams. Critical ocean studies and blue humanities both are generative of lively and much needed, but almost always sea-focused conversations (Elizabeth DeLoughrey, "Submarine Futures of the Anthropocene," *Comparative Literature* 69, no. 1 (2017): 32–44; Steve Mentz, "Blue Humanities," in *Posthuman Glossary*, ed. by Rosi Braidotti and Maria Hlavajova (London: Bloomsbury, 2018), 69–72; John R. Gillis,

"The Blue Humanities," *The Humanities* 34, no. 3 (2013); Elizabeth DeLoughrey, "Toward a Critical Ocean Studies for the Anthropocene," *English Language Notes* 57, no. 1 (2019): 21–36. As such, my work is very much in conservation with, while also exceeding what scholars call an "oceanic turn," or, as Stefan Helmreich provokes, an "oceanic churn" (Helmreich, *A Book of Waves*).

13. Dilip da Cunha, *The Invention of Rivers: Alexander's Eye and Ganga's Descent* (Philadelphia: University of Pennsylvania Press, 2019).

14. da Cunha, *The Invention of Rivers*, ix.

15. Barad, *Meeting the Universe Halfway*.

16. Haraway, "Situated Knowledges: The Science Question in Feminism and the Privilege of Partial Perspective," 575–599.

17. da Cunha, *The Invention of Rivers*, 1.

18. da Cunha, *The Invention of Rivers*, xi.

19. Deleuze and Guattari's lines of flight are a canonical example from Western posthuman philosophy, where lines map emergent possibilities of escape from dominating terrains of power (though transformation is never guaranteed). Yet even here, the terms of settler colonialism tend to be reinscribed, as Jodi Byrd has argued. In the words of Tiffany Lethabo King, "White posthumanism and its flows and lines of flight are made possible through Native death." Gilles Deleuze and Félix Guattari, *A Thousand Plateaus: Capitalism and Schizophrenia* (Minneapolis, MN: Continuum, 2004); Jodi A. Byrd, *The Transit of Empire: Indigenous Critiques of Colonialism*, NED—New edition. (Minneapolis, MN: University of Minnesota Press, 2011), xxxix; Tiffany Lethabo King, *The Black Shoals: Offshore Formations of Black and Native Studies* (Durham, NC: Duke University Press, 2019), 100.

20. Audre Lorde, calling for the necessity of difference in community, writes "As women, we have been taught either to ignore our differences or to view them as causes for separation and suspicion rather than as forces for change." Audre Lorde, "The Master's Tools Will Never Dismantle the Master's House," in *This Bridge Called My Back, Fortieth Anniversary Edition*, ed. by Cherríe Moraga and Gloria Anzaldúa (SUNY Press, 2021), 95; Barad, "Diffracting Diffraction: Cutting Together-Apart."

21. Alexander the Great was a student of Aristotle, whose classical articulation of the Greek elements not only refined the four Western elements but also reorganized them spatially. Where Thales, whom Aristotle deemed the founder of Greek and natural philosophy, saw water as the foundational substance from which all others originated, Aristotle privileged earth, placing it at the center-bottom of his cosmos where it was surrounded by or overlain with water. Aristotle did, however, retain Thales's sense of elements as both substances and processes. The four elements could transform into one another through changes in quality: Water heated transformed into air; water dried transformed into earth. The corresponding world map, a central circular landmass surrounded by an outer sea, is quite literally terracentric. da Cunha, *The Invention of Rivers*, 22; Irene J. Klaver, "Meander(ing) Multiplicity," in *Water*

Scarcity, Security and Democracy: A Mediterranean Mosaic, ed. by Francesca de Châtel, Gail Holst-Warhaft, and Tammo Steenhuis (Global Water Partnership Mediterranean, Cornell University, and Atkinson Center for a Sustainable Future, 2014), 39; Jamie Linton, *What Is Water?: The History of a Modern Abstraction* (Vancouver, BC: UBC Press, 2010), 76.

22. In focusing on exclusions enacted by the line, I take up feminist projects of asking "how difference can be done differently" Barad, "Diffracting Diffraction: Cutting Together-Apart," 168–187. As Melody Jue writes of terrestrial bias, the problem is not embodying a land-based perspective but simultaneously treating it as a neutral perspective while imposing it elsewhere. "I imagine 'terrestrial bias' as a necessary partial perspective—one that, once recognized, erodes the dream of a master language that would be totally objective, distant, and adequate to articulating and describing the world in its entirety." Jue, *Wild Blue Media*, 11. Where terrestrial bias calls out land-based views to insist they are always views from somewhere, terracentrism calls out worlds built of elemental domination that privilege (bounded) ground as ontological, epistemological, political, and ethical foundation.

23. da Cunha, *The Invention of Rivers*. Irene Klaver, "Radical Water," in *Hydrohumanities: Water Discourse and Environmental Futures*, ed. by Kim De Wolff, Rina C. Faletti, and Ignacio López-Calvo (Oakland, CA: University of California Press, 2022), 64.

24. Slat, "How the Oceans Can Clean Themselves."

25. Hawaii Statewide GIS Program. "Ahupuaa." (2024). https://geoportal.hawaii.gov/datasets /07624815fc7d42d4b23c527d20ad2f58_1/explore?location=18.960920%2C-155 .676299%2C12.17

26. Noa K. Lincoln, Mehana Blaich Vaughan, and Natalie Kurashima, "Hawai'i," in *Islands & Cultures*, ed. by Kamanamaikalani Beamer, Te Maire Tau, and Peter Morrison Vitousek (New Haven, CT; London: Yale University Press, 2022), 35–75, 42; Candace Fujikane, *Mapping Abundance for a Planetary Future* (Durham, NC: Duke University Press, 2021), 4.

27. Lincoln et al., "Hawai'i," 35–75, 36.

28. Lincoln et al., "Hawai'i," 64.

29. Karin Amimoto Ingersoll, *Waves of Knowing: A Seascape Epistemology* (Durham, NC: Duke University Press, 2016).

30. Fujikane, *Mapping Abundance for a Planetary Future*, 19.

31. Manulani Aluli Meyer et al., "Special Plenary: Aloha 'Āina: Hawaiian Knowledge Today," *Society for the Social Study of Science*, 2023.

32. Fujikane, *Mapping Abundance for a Planetary Future*, 18.

33. Linton, *What Is Water?*

34. As Jamie Linton points out in his history of modern water, even Robert Boyle, credited with originating the modern scientific experiment in the seventeenth century, held that water was a mutable element. Linton, *What Is Water?*, 76–77.

35. In doing so, Lavoisier seriously upended several thousand years of elemental tradition. Water's place as a discrete element is so deeply steeped in Western culture that the revelation of its hydrogen-and-oxygen component form took modern chemistry very much by surprise.

36. Klaver, "Meander(ing) Multiplicity," 38–47.

37. Nicole Starosielski, "The Elements of Media Studies," *Media + Environment* 1, no. 1 (2019). See also Dimitri Papadopoulos, María Puig de la Bellacasa, and Natasha Myers. *Reactivating Elements: Chemistry, Ecology, Practice* (Durham, NC: Duke University Press, 2022).

38. Jue and Ruiz, *Saturation*; Astrida Neimanis, *Bodies of Water: Posthuman Feminist Phenomenology* (London: Bloomsbury Publishing Plc, 2017).

39. Yuriko Furuhata, "Of Dragons and Geoengineering: Rethinking Elemental Media," *Media + Environment* 1, no. 1 (2019).

40. In New Zealand, where mountains and a river have been awarded legal personhood status, the very existence of a line separating land from sea explodes from a philosophical question into a site of practical politics. As geographer of oceans Kate Sammler describes in her account of protests against undersea mineral extraction, at stake is sovereignty and control over ocean resources. With legal personhood awarded to Mount Taranaki, a volcanic peak at the cusp of the Pacific Ocean, indigenous Māori and the New Zealand government's shared guardianship is fraught with a fundamental disagreement about where, exactly, the mountain ends. For the settler colonial government grounded in western ontology, the mountain stops at the shore, at the line dividing land from seawater, which demarcates the seabed as state resource. For Māori, whose worldview and associated articulation of responsibility constitute landscape and seascape as an interrelated whole, the mountain extends to the sea floor. The same seabed whose rich mineral deposits have been eye-marked by industry as the next frontier for large-scale resource extraction. A situation that points to the enduring limitations of visions of justice rooted in "inclusion" where it very much matters who is being included in what and on whose terms. Katherine Sammler, "Kauri and the Whale: Oceanic Matter and Meaning in New Zealand," in *Blue Legalities: The Life and Laws of the Sea*, ed. by Irus Braverman and Elizabeth R. Johnson (Durham, NC: Duke University Press, 2020), 63–84.

41. European settlers overwhelmingly adapt land to plants rather than vice versa. Amitav Ghosh, *The Nutmeg's Curse: Parables for a Planet in Crisis* (Chicago: The University of Chicago Press, 2021); Robin Wall Kimmerer, *Braiding Sweetgrass: Indigenous Wisdom, Scientific Knowledge and the Teachings of Plants* (London: Penguin Books, 2020).

42. On the constitutive mediations and fluid challenges of modeling oceans in general and the Pacific Ocean in particular see Sebastian Vehlken et al., "Introduction: Modeling the Pacific Ocean," *Media + Environment: Modeling the Pacific Ocean* 3, no. 2 (2021).

43. Day and Shaw, "Patterns in the Abundance of Pelagic Plastic and Tar in the North Pacific Ocean, 1976–1985,": 311–316, 261.

44. Eric L. Mills, *The Fluid Envelope of Our Planet: How the Study of Ocean Currents Became a Science* (Toronto: University of Toronto Press, 2009), 17.

45. Carmel Finley, *All the Fish in the Sea: Maximum Sustainable Yield and the Failure of Fisheries Management* (Chicago; London: The University of Chicago Press, 2011), xii.

46. Floatee exploits are intriguingly chronicled in *Moby-Duck*; and the science in *Flotsametircs* respectively. Donovan Hohn, *Moby-Duck: The True Story of 28,800 Bath Toys Lost at Sea and of the Beachcombers, Oceanographers, Environmentalists, and Fools, Including the Author, Who Went in Search of Them* (Viking, 2011); Curtis Ebbesmeyer and Eric Scigliano, *Flotsametrics and the Floating World* (HarperCollins Publishers, 2010).

47. Laurent Lebreton et al., "Evidence That the Great Pacific Garbage Patch Is Rapidly Accumulating Plastic," *Scientific Reports* 8, no. 1 (2018): 4666–15. https://doi.org/10.1038/s41598-018-22939-w.

48. Geoffrey C. Bowker and Susan Leigh Star, *Sorting Things Out: Classification and its Consequences* (Cambridge, MA: The MIT Press, 1999).

49. Brian Russell Roberts, *Borderwaters: Amid the Archipelagic States of America* (Durham, NC: Duke University Press, 2021).

50. Lebreton et al., "Evidence That the Great Pacific Garbage Patch Is Rapidly Accumulating Plastic," 15.

51. Liboiron, *Pollution Is Colonialism*.

52. Alexander Mawyer, "Floating Islands, Frontiers, and Other Boundary Objects on the Edge of Oceania's Futurity," *Pacific Affairs* 94, no. 1 (2021), 136.

53. Roberts, *Borderwaters*.

54. Teresia Teaiwa, "To Island," in *A World of Islands: An Island Studies Reader*, ed. by Godfrey Baldacchino (Charlottetown, PEI: University of Prince Edward Island and Agenda Academic, 2007), 514. Baldacchino, Godfrey, and Eric Clark. "Guest Editorial Introduction: Islanding Cultural Geographies." *Cultural Geographies* 20, no. 2 (2013): 129–134.

55. As Hau'ofa writes: "Continental men-Europeans and Americans-drew imaginary lines across the sea, making the colonial boundaries that confined ocean peoples to tiny spaces for the first time. These boundaries today define the island states and territories of the Pacific. I have just used the term ocean peoples because our ancestors, who had lived in the Pacific for over two thousand years, viewed their world as 'a sea of islands' rather than as 'islands in the sea.'" Epeli Hau'ofa, "Our Sea of Islands," *The Contemporary Pacific* 6, no. 1 (1994): 147–161; Elizabeth DeLoughrey, "The Myth of Isolates: Ecosystem Ecologies in the Nuclear Pacific," *Cultural Geographies* 20, no. 2 (2013): 167–184.

56. Michelle Murphy, "Alterlife and Decolonial Chemical Relations," *Cultural Anthropology* 32, no. 4 (2017): 494–503.

57. Murphy, "Alterlife and Decolonial Chemical Relations," 258.

58. Traci Brynne Voyles, *Wastelanding: Legacies of Uranium Mining in Navajo Country* (Minneapolis, MN; London: University of Minnesota Press, 2015), 7.

59. Joshua Reno, *Military Waste: The Unexpected Consequences of Permanent War Readiness* (Oakland, CA: University of California Press, 2019), 183.

60. John R. Gillis, *Islands of the Mind: How the Human Imagination Created the Atlantic World* (New York: Palgrave Macmillan, 2004), 2.

61. Philip Steinberg, *The Social Construction of the Ocean* (Cambridge, UK: Cambridge University Press, 2001), 105.

62. Steinberg, *The Social Construction of the Ocean*; Helen M. Rozwadowski, *Vast Expanses: A History of the Oceans* (London: Reaktion Books, 2018); Helen M. Rozwadowski, "Arthur C. Clarke and the Limitations of the Ocean as a Frontier," *Environmental History* 17, no. 3 (2012): 578–602; David Armitage and Alison Bashford. *Pacific Histories: Ocean, Land, People* (Basingstoke, UK: Palgrave Macmillan, 2014), 15.

63. Elizabeth DeLoughrey, *Allegories of the Anthropocene* (Durham, NC: Duke University Press, 2019), 112.

64. Dan Elbert Clark, "Manifest Destiny and the Pacific," *Pacific Historical Review* 1, no. 1 (1932): 1–17.

65. Frederick J. Turner, "The Significance of the Frontier in American History," January 1, 1893. Available from https://www.historians.org/resource/the-significance-of-the-frontier-in-american-history/.

66. William Cronon, "The Trouble with Wilderness; Or, Getting Back to the Wrong Nature," *Environmental History* 1, no. 1 (1996): 7–28.

67. Carolyn Finney, *Black Faces, White Spaces: Reimagining the Relationship of African Americans to the Great Outdoors* (Chapel Hill, NC: University of North Carolina Press, 2014).

68. Turner, "The Significance of the Frontier in American History."

69. Reno, *Military Waste*. US guano-island expansion was also driven in part by competition with the British amid a Peruvian guano trade monopoly. Christina Duffy Burnett, "The Edges of Empire and the Limits of Sovereignty: American Guano Islands," *American Quarterly* 57, no. 3 (2005): 779–803.

70. Burnett, "The Edges of Empire and the Limits of Sovereignty," 780.

71. Burnett, "The Edges of Empire and the Limits of Sovereignty," 788; Jennifer C. James, "'Buried in Guano': Race, Labor, and Sustainability," *American Literary History* 24, no. 1 (2012): 115–142.

72. Burnett, "The Edges of Empire and the Limits of Sovereignty," 798.

73. Reno, *Military Waste*, 183.

74. Kamanamaikalani Beamer, *No Mākou ka Mana: Liberating the Nation* (Honolulu: Native Books, 2014).

75. Ashanti Shih, "The Most Perfect Natural Laboratory in the World: Making and Knowing Hawaii National Park—Ashanti Shih, 2019," *History of Science* 57, no. 4 (2019): 493–517.

76. John F. Kennedy, "The New Frontier," *Democratic National Convention.* July 15, 1960. https://www.presidency.ucsb.edu/documents/address-accepting-the-democratic-nomination-for-president-the-memorial-coliseum-los.

77. Kennedy, "The New Frontier."

78. Rozwadowski, *Vast Expanses*, 167. The frontier is of sea and space converging in Stefan Helmreich's *Alien Ocean*.

79. Elizabeth DeLoughrey, "The Myth of Isolates: Ecosystem Ecologies in the Nuclear Pacific," 167–184.

80. DeLoughrey, "The Myth of Isolates."

81. DeLoughrey, "The Myth of Isolates;" Joseph Masco, "The Artificial World," in *Reactivating Elements: Chemistry, Ecology, Practice*, ed. by Dimitris Papadopoulos, María Puig de la Bellacasa, and Natasha Myers (Durham, DC: Duke University Press, 2022), 131–150.

82. For an expansive study of plastic pollution as a continuation of waste colonialism in the Pacific see: Sascha Fuller, Tina Ngata, Stephanie B. Borrelle, and Trisia Farrelly. "Plastics Pollution as Waste Colonialism in Te Moananui." *Journal of Political Ecology* 29, no. 1 (2022): 534–560.

83. Elodie Fache et al., "Introduction: The New Scramble for the Pacific: A Frontier Approach," *Pacific Affairs* 94, no. 1 (2021): 57–76.

84. Fache et al., "Introduction: The New Scramble for the Pacific: A Frontier Approach," 75. Synthetic frontiers draw further emphasis to the violence of imposed Western ways of knowing by focusing in detail on the material practices of Western science and technology and by refusing to separate epistemology from ontology and responsibility.

85. Fache et al., "Introduction: The New Scramble for the Pacific: A Frontier Approach," 57–76, 66.

86. As Kroll writes in his "wilderness" history of American oceans, in the early twentieth century scientific management practices caught up with conserving ocean resources for industrial capitalism. Gary Kroll, *America's Ocean Wilderness: A Cultural History of Twentieth-Century Exploration* (La Vergne, KS: University Press of Kansas, 2008).

87. Reno, *Military Waste*. In other manifestations, "military to wildlife conversion" becomes an excuse for minimal remediation while military activities move on to fresher lands. Thom van Dooren, *A World in a Shell: Snail Stories for a Time of Extinctions* (Cambridge, MA: The MIT Press, 2022), 168–69.

88. Reno, *Military Waste*, 195.

89. Maite Zubiaurre, *Talking Trash: Cultural Uses of Waste* (Nashville, TN: Vanderbilt University Press, 2019), 193–195.

90. Harris and Morazzo, *Great Pacific Volume 1: Trashed!*; Harris and Morazzo, *Great Pacific Volume 2: Nation Building*; Harris and Morazzo, *Great Pacific Volume 3: Big Game Hunters*.

91. Here the many references to a garbage patch "twice the size of Texas" exceeds the truism that "that everything is bigger in Texas," cementing ocean plastic in relation with geopolitical state borders and the fossil fuel industry.

92. This final move is not the end of the synthetic frontier. Rather, the linear narrative of technological progress overcoming human greed is itself a perpetuation of the synthetic frontier. The good-feeling transitions of green energy, including the building of green energy islands, are instigating further investment in plastic and petrochemical production, rather than petrocaptialist undoing. May Ee Wong, "Refiguring the 'Renewable Energy Island' as Future Environment," Society for the Social Study of Science, 2023.

93. WHIM Architecture, "Recycled Island," August 21, available from http://www.whim.nl /Recycledisland.html.

94. Petráková "The 8th Continent," [cited 2023]. Available from https://lenkapetrakova.com /the-8th-continent-info.

95. Much of what is on Kamilo Beach, as in the garbage patch, is the stuff of industry and even science: all manner of fishing gear, lost nets, crates; plastic tubes escaped from aquaculture that I learn to call oyster spacers; the remains of intentionally launched weather balloons and tracker buoys. If you look closely among the synthetic sand, there are white, beadlike spheres too regular to be fragments of larger items. These are preproduction plastic pellets, nicknamed "nurdles," escaped industrial materials.

96. Eleanor Gibson, "The Ocean Cleanup Suspended as Device Breaks Down in Pacific Ocean," *Dezeen*, January 7, 2019.

REORIENTATIONS

1. The *Sea Dragon* was built for the Global Challenge Race in 2000, designed to travel around the world against the prevailing winds or the "wrong way" (Pangaea Exploration 2025). It is an incredibly seaworthy vessel—though it was made for safety and speed rather than research.

CHAPTER 2

1. ChatGPT response to "Write me a 1,000-word essay about the plastic trash island in the Pacific. List five sources at the end." ChatGPT, Open AI, January 5, 2024. https://chat.openai .com/share/f66de41e-539c-42fa-ac28-e45c7e9284e0.

2. All of the sources ChatGPT provided at my request actually exist and are not made up. Though none of them provide evidence of a trash island, they did include the Lebreton et al. 2018 article in *Nature* whose visualizations I argue are easily read as an island in chapter 1, and a 2014 article by Marcus Eriksen, my expedition watchmate aboard the *Sea Dragon* in 2011. Lebreton et al., "Evidence that the Great Pacific Garbage Patch Is Rapidly Accumulating Plastic"; Marcus Eriksen et al., "Plastic Pollution in the World's Oceans: More Than 5 Trillion Plastic Pieces Weighing over 250,000 Tons Afloat at Sea," *PLoS ONE* 9, no. 12 (2014).

3. Berton, "Continent-Size Toxic Stew of Plastic Trash Fouling Swath of Pacific Ocean"; *ABC News*, "Hidden, a 3.5 Million Trash Heap Lies in the Ocean"; Daily Mail, "Rubbish Dump Found Floating in Pacific Ocean Is Twice the Size of America," *Daily Mail.com*, February 6, 2008; Hoshaw, "Afloat in the Ocean, Expanding Islands of Trash"; Kostigen, *The World's Largest Dump: The Great Pacific Garbage Patch*; Pravda, "'Trash Island' Discovered in the Pacific Ocean," *English Pravda Online*, February 24, 2004; Bryan Walsh, "The Truth About Plastic," *Time*, July 10, 2008; WHIM Architecture, "Recycled Island."

4. Annalee Newitz, "Lies You've Been Told About the Pacific Garbage Patch," *Gizmodo*, 2012.

5. Pravda, "'Trash Island' discovered in the Pacific Ocean."

6. At the time, the thumbnail still linked to the *Pravda* article in question, but the image had been removed.

7. The allure of debunkery is not the domain of natural scientists and realists alone. Latour introduced matters of concern along with a prescient warning that the tools of STS, especially social construction, were being wielded as weapons by extremists bent on undermining scientific consensus for political gain. Presenting scientific facts as always social and political has the potential to make them vulnerable, as the assertion that all facts—good and bad, true and false alike—are "made" is conflated with a kind of extreme relativism where all facts are equally "made up." Claims that facts are carefully negotiated, curated, and tended were taken up to fuel arguments that facts are whatever you want them to be, or heated accusations that STS scholars did not believe in reality. Matters of concern is Latour's attempt (with a nod to Donna Haraway) "to devise another powerful descriptive tool that deals this time with matters of concern and whose import then will no longer be to debunk but to protect and to care" (232). Bruno Latour, "Why Has Critique Run Out of Steam? From Matters of Fact to Matters of Concern," *Critical Inquiry* 30, no. 2 (2004), 225–248; Bruno Latour, *What Is the Style of Matters of Concern? Two Lectures in Empirical Philosophy* (Koninklijke Van Gorcum, 2008).

8. Maria Puig de la Bellacasa, *Matters of Care: Speculative Ethics in More Than Human Worlds* (University of Minnesota Press, 2017), 61.

9. Karen Barad, "Posthumanist Performativity: Toward an Understanding of How Matter Comes to Matter," *Signs: Journal of Women in Culture and Society* 28, no. 3 (2003), 801–831.

10. There was no word for the Pacific as a whole, which arguably did not even exist as such, until "discovered" and named by Europeans. Damon Salesa, "The Pacific in Indigenous Time," in *Pacific Histories: Ocean, Land, People*, ed. by David Armitage and Alison Bashford (Palgrave Macmillan, 2014), 41.

11. As Amitav Ghosh brings to light in *The Nutmeg's Curse*, the Dutch pursuit of not just spices, but a spice-trade monopoly energized a "campaign of extermination" waged at both island peoples and trees alike (74). Islands of connection are "terraformed" by European relations that grounded value in resource scarcity at the expense of island abundance. Ghosh, *The Nutmeg's Curse*,

12. Sujit Sivasundaram, "Science," in *Pacific Histories: Ocean, Land, People*, ed. by David Armitage and Alison Bashford (Palgrave Macmillan, 2014), 244–247.

13. Sivasundaram, "Science," 237–261.

14. Sivasundaram, "Science." Lost continents continue to populate headlines, as they continue to be discovered by geoscientists searching geological pasts and underneath the sea. Maria Seton et al., "What Are Lost Continents, and Why Are We Discovering So Many?" *The Conversation*, November 24, 2019

15. Barad, *Meeting the Universe Halfway*; Donna Haraway, *When Species Meet* (Minneapolis, MN: University of Minnesota Press, 2008); Latour, *We Have Never Been Modern*; Puig de la Bellacasa, *Matters of Care*; Bruno Latour, *Pasteurization of France* (Harvard University Press, 1993); Donna Haraway, "A Manifesto for Cyborgs: Science, Technology, and Socialist Feminism in the 1980s, " in *The Haraway Reader* (Routledge, 2004), 7–45.

16. Barad, *Meeting the Universe Halfway*, 3.

17. Barad, *Meeting the Universe Halfway*, 3–6.

18. Zoe Todd, "An Indigenous Feminist's Take on the Ontological Turn: 'Ontology' Is Just Another Word for Colonialism," *Journal of Historical Sociology* 29, no. 1 (2016), 4–22; Vanessa Watts, "Indigenous Place-Thought Agency Amongst Humans and Non-Humans (First Woman and Sky Woman go on a European World Tour!)," *Decolonization* 2, no. 1 (2013); Vincente Diaz, "No Island Is an Island," in *Native Studies Keywords*, ed. by Stephanie Nohelani Teves, Andrea Smith, and Michelle Raheja (Chicago: The University of Arizona Press, 2015), 90–107.

19. As holds for many critiques of new materialism and posthumanism especially from Indigenous, Black, decolonial, and feminist scholars. Todd, "An Indigenous Feminist's Take on the Ontological Turn: 'Ontology' Is Just Another Word for Colonialism," 4–22; Puig de la Bellacasa, *Matters of Care;* Jackson, *Becoming Human*; Zakiyyah Iman Jackson, "Outer Worlds: The Persistence of Race in Movement "Beyond the Human," *GLQ* 21, no. 2 (2015), 215–218.

20. Puig de la Bellacasa, *Matters of Care*.

21. As ocean scholars turn to the multidimensional depths of "wet" ontologies, STS scholars explore "radical geometric reflexivity" that exceeds both horizontal and vertical analysis scaffolded by a volumetric grid to consider shifting elemental states of matter. Turn to the volumetric grid. Steinberg and Peters, "Wet Ontologies, Fluid Spaces: Giving Depth to Volume Through Oceanic Thinking," 247–264; Tone Walford and Lydia Gibson, "The Shapes of Things: Revising Geometries and Acknowledging Forms of/in/Through Territories, Terrains, and Technologies," Society for the Social Study of Science, 2023; Barad, "Diffracting Diffraction: Cutting Together-Apart," 168–187.

22. Ramon Knoester, Personal Correspondence with the Author, 2014.

23. Ramon Knoester, "Recycled Living Spaces," *The Journal of Ocean Technology* 11, no. 2 (2016).

24. Cian Luanaigh, "Recycled Island: Plastic Fantastic?" *The Guardian*, August 7, 2010.

25. The life-and-death entanglements of Laysan albatross with plastic have garnered massive public attention in no small part due to the work of artist Chris Jordan, especially the film *ALBATROSS*. Chris Jordan, *ALBATROSS*, 2017. For further discussion of albatross and the ethics of making nonhuman trauma visible, see chapter 3.

26. Luanaigh, "Recycled Island: Plastic Fantastic?"

27. Ramon Knoester, Personal Correspondence with the Author, 2014.

28. With gratitude to Irene Klaver for pointing out the parallels with tulip fields—so many of which are themselves ground reclaimed from the sea.

29. Lizabeth Cohen, *A Consumer's Republic: The Politics of Mass Consumption in Postwar America* (New York: Vintage, 2003).

30. Ramon Knoester, Personal Correspondence with the Author, 2014.

31. Bryan Nelson, "Hawaii-Sized Recycled Island to Be Built from Ocean Garbage Patch," *Mother Nature Network*, April 1, 2011.

32. Ariel Schwartz, "Electrolux Turning Plastic from the Ocean into Vacuum Cleaners," *Fast Company*, 2010.

33. Miriam Goldstein, interview with the author, 2012.

34. Maya J. Goldenberg, *Vaccine Hesitancy: Public Trust, Expertise, and the War on Science* (Pittsburgh: University of Pittsburgh Press, 2021).

35. Angelicque White, interview with the author, 2012.

36. National Oceanic and Atmospheric Administration, "What Are Marine Microbes?," [cited 2024]. Available from https://oceanexplorer.noaa.gov/facts/marinemicrobes.html.

37. Angelicque White, interview with the author, 2012.

38. Angel White, "Oceanic "Garbage Patch" Not Nearly as Big as Portrayed in Media," January 04, Available from https://web.archive.org/web/20140919020832/http://oregonstate.edu/ua/ncs/archives/2011/jan/oceanic-%E2%80%9Cgarbage-patch%E2%80%9D-not-nearly-big-portrayed-media.

39. White, "Oceanic "Garbage Patch" Not Nearly as Big as Portrayed in Media."

40. Naomi Oreskes and Erik M. Conway, *Merchants of Doubt: How a Handful of Scientists Obscured the Truth on Issues from Tobacco Smoke to Global Warming* (New York: Bloomsbury Press, 2010).

41. Miriam Goldstein, in contrast, was surprised by how easy plastic was to find at sea and ended up on board with there being an ocean plastic crisis worthy of attention. However, she most definitely agreed with White that action based on anything other than scientific facts was dangerous to the cause. As Goldstein explained on the SEAPLEX website: "Widespread misinformation, as is so common regarding plastic in the North Pacific, serves no one—not activists trying to ban plastic bags, not plastic manufacturers trying to develop ocean-degradable products, not groups developing methods to stop plastic pollution. Our role as scientists is

to find out truths about the world, and to interpret and explain them." Miriam Goldstein, "Does the 'Great Pacific Garbage Patch' Exist?," January 10, [cited 2014]. Available from https://web.archive.org/web/20130727212124/http://seaplexscience.com/2011/01/10/does -the-great-pacific-garbage-patch-exist/.

42. Toby Murcott, "Science Journalism: Toppling the Priesthood," *Nature* 459 (2009), 1054– 1055; Justin Matthew Wren Lewis et al., "A Compromised Fourth Estate? UK News Journalism, Public Relations and News Sources," *Journalism Studies* 9, no. 1 (2008), 1–20.

43. Marc Gunther, "In Defense of the Plastic Bag," December 22, [cited 2024]. Available from https://www.greenbiz.com/article/defense-plastic-bag.

44. Barry E. DiGregorio, "Tracking Plastic in the Oceans," *EARTH Magazine*, 2012.

45. Jeanne Gallagher, interview with the author, 2012.

46. Donovan Hohn, "Sea of Trash," *The New York Times*, June 22, 2008.

47. Moore and Phillips, *Plastic Ocean*.

48. Katie Allen, interview with the author, 2012.

49. The Bryant Park Project, "Garbage Mass Is Growing in the Pacific," *National Public Radio*, March 26, 2008.

50. *ABC News*, "Pacific Ocean to Receive Plastic Island," June 30, 2010; Nelson, *Hawaii-Sized Recycled Island to Be Built from Ocean Garbage Patch*; Stephen Messenger, "'Recycled Island' Turns Ocean Plastic into a Paradise," June 28, [cited 2024]. Available from https://www .treehugger.com/recycled-island-turns-ocean-plastic-into-a-paradise-4858242.

51. Miriam Goldstein, "'Recycled Island' Not a Cure for Plastic Trash in Ocean," *Deep Sea News*, August 9, 2010.

52. Ramon Knoester, Personal Correspondence with the author, 2014.

53. "CLEAR RIVERS | For a Plastic Free Sea!" [cited 2024]. Available from https://www .clearrivers.eu/.

54. Lynne Myers, "The 8th Continent by Lenka Petráková is a Floating Station Concept to Clean Up Our Oceans," *Designboom*, January 6, 2021.

55. Myers, "The 8th Continent."

56. Melody Jue, "Floating Architectures," in *The Routledge Companion to Media and Risk*, ed. by Bishnupriya Ghosh and Bhaskar Sarkar (Routledge, 2020a), 316.

57. Here islands become laboratories for libertarian social and political science experiments. DeLoughrey, *Allegories of the Anthropocene*, 145.

58. The Seasteading Institute, "About—The Seasteading Institute," [cited 2024]. Available from https://www.seasteading.org/about/.

59. A headline, I learn in my island-origin tracing attempts, that was also very much the product of entangled relations, not individual choice alone. The author, Lindsey Hoshaw, a journalism

graduate student, who had just returned from a garbage patch expedition aboard the *Alguita* with Moore and was about to publish her very first article in the prestigious paper. The short story of the headline is a back and forth with an insistent editor who ended up running the headline despite repeated attempts to clarify that ocean plastic pollution was not in the form of an island. When I interviewed Hoshaw, she had heard from ocean scientists, and been in conversation with Goldstein specifically and was repentant about the headline. At the same time, Howshaw's practical understanding of communication intriguingly exceed the scientific standards of linear, accurate transmission of matters of fact (or corporate media demands of directing eyes to advertisements) to exemplify the multiple meanings of "matter" in action. In conceptualizing her article, Hoshaw's question was not simply, "How do I describe the physical form of the garbage patch?," but at the same time "So what?" As she continued: "It's so easy to say our oceans are polluted, our lands are polluted, our roads are polluted, so what. Why does this matter and why should I care if I'm someone living in the middle of the country, that thousands of miles away that there's some popsicle sticks floating in the ocean. So I really wanted to try to bring it home, and talk about why should this matter, why should you care about this." Our conversation pointed to material-ethical tensions between accurately conveying physical form and accurately conveying the importance of it as a problem: "It's really a challenge, because you want people to be galvanized to action, you want people to care, but you don't want to misrepresent it by saying it's the eight continent and it's going to kill us all and, you know, people can colonize it." In Hoshaw's studied and practical experience of communication, the reality of the physical form of plastic at sea in scientific terms and the reality that her readers should be concerned about it did not so easily mesh. "Because if you just jump in the water and take a photo of what you see, you may see a couple confetti-sized fragments, but you might not see anything and how do you use that to convince people that this garbage patch is real? You don't." (Lindsey Hoshaw, interview with the author, 2013)

60. Charles J. Moore, "Choking the Oceans with Plastic," *The New York Times*, August 25, 2014.

61. Charles J. Moore, writing from aboard the research vessel *Alguita*. "Blog 14, July 13, 2014."

62. iADYS "Jellyfishbot a Solution for Water Cleaning, Depollution, & Preservation," [cited 2024]. Available from https://www.iadys.com/jellyfishbot/.

63. Puig de la Bellacasa, *Matters of Care*, 32.

64. Helmreich, *A Book of Waves*, xvii.

65. Michelle Murphy, "Studying Unformed Objects: Deviation," *Fieldsights* (2013).

66. Puig de la Bellacasa, *Matters of Care*, 61.

67. Frederick, "Ninety-Nine Percent of Ocean Plastic Has Gone Missing,"

68. Andrés Cózar et al., "Plastic Debris in the Open Ocean," *Proceedings of the National Academy of Sciences* 111, no. 28 (2014), 10239–10244.

CHAPTER 3

1. Erik R. Zettler et al., "Life in the "Plastisphere": Microbial Communities on Plastic Marine Debris," *Environmental Science & Technology* 47, no. 13 (2013), 7137–7146.

2. Linsey E. Haram et al., "A Plasticene Lexicon," *Marine Pollution Bulletin* 150, (2020), 110714.; Zettler et al., "Life in the "Plastisphere": Microbial Communities on Plastic Marine Debris," 7137–7146.

3. Matthias C. Rillig et al., "The Soil Plastisphere," *Nature Reviews Microbiology* 22 (2024), 64–74.

4. Zettler et al., "Life in the "Plastisphere": Microbial Communities on Plastic Marine Debris," 7137–7146.

5. Miriam C. Goldstein et al., "Relationship of Diversity and Habitat Area in North Pacific Plastic-Associated Rafting Communities," *Marine Biology* 161 (2014), 1441–1453. The authors argue that the presence of ocean plastic allows "islands" of substrate-associated organisms to persist in an otherwise unsuitable habitat."

6. Murray R. Gregory, "Environmental Implications of Plastic Debris in Marine Settings—Entanglement, Ingestion, Smothering, Hangers-on, Hitch-Hiking and Alien Invasions," *Philosophy Transactions of the Royal Society B* 364, no. 1526 (2009), 2013–2025; Henry S. Carson et al., "The Plastic-Associated Microorganisms of the North Pacific Gyre," *Marine Pollution Bulletin* 75, no. 1–2 (2013), 126–132.

7. Steve Olson, *Evolution in Hawaii: A Supplement to Teaching About Evolution and the Nature of Science* (Washington, DC: National Academies Press, 2004).

8. Zettler et al., "Life in the "Plastisphere": Microbial Communities on Plastic Marine Debris," 7137–7146; Robyn J. Wright et al., "Marine Plastic Debris: A New Surface for Microbial Colonization, " *Environmental Science & Technology* 54, no. 19 (2020), 11657–11672.
    For a more recent overview, see Linda A. Amaral-Zettler, "Colonization of Plastic Marine Debris: The Known, the Unknown, and the Unknowable," in *Plastics and the Ocean: Origin, Characterization, Fate, and Impacts*, ed. by Anthony L. Andrady (John Wiley & Sons, Incorporated, 2022).

9. Linda A. Amaral-Zettler et al., "Ecology of the Plastisphere," *Nat Rev Microbiol* 18, no. 3 (2020), 139–151.

10. Fujikane, *Mapping Abundance for a Planetary Future*, 23–24.

11. Zettler et al., "Life in the "Plastisphere": Microbial Communities on Plastic Marine Debris," 7137–7146. In this way, the plastic (once alive oil, not alive plastic, host to new life) comes to embody the figure of contemporary power that Elizabeth Povinelli calls "the Desert" "the space where life was, is not now, but could be if knowledges, techniques, and resources where properly managed." Elizabeth A. Povinelli, *Geontologies: A Requiem to Late Liberalism* (Duke University Press, 2016), 16.

12. Delving into the emergence of the term "biosphere" surfaces uneasy divides between substances and spaces, even challenging that very distinction. In its nineteenth-century origins, biosphere referred to the zone where life occurred on earth. The term was first proposed by Eduard Suess, an Austrian geologist whose book describing the origins of the Alps, introduces biosphere among a set of spatial concepts for describing the planet as a whole. In doing so,

Suess also earns credit with expanding geology as the study of earth's layers from a local to a planetary scale, the very scale upon which epochal markers are now judged. Popularized by other scholars in the twentieth century, biosphere came to mean the layer of life itself, rather than a place defined in relation with the presence of life, though competing meanings persist as it is still sometimes used to gather organisms along with their media or habitat. Such competing meanings persist and proliferate, not only with biosphere, but with geosphere, which is sometimes used interchangeably with lithosphere as the layer of solid rock, but elsewhere encompassing an entire planetary system, including lively relations. R. J. Huggett, "Ecosphere, Biosphere, or Gaia? What to Call the Global Ecosystem," *Global Ecology and Biogeography* 8 (2001), 425–431.

13. Robyn J. Wright et al., "Marine Plastic Debris: A New Surface for Microbial Colonization," 11657–11672; Amaral-Zettler, "Colonization of Plastic Marine Debris: The Known, the Unknown, and the Unknowable," in Amaral-Zettler et al., "Ecology of the Plastisphere," 139–151; Zettler et al., "Life in the "Plastisphere": Microbial Communities on Plastic Marine Debris," 7137–7146.

14. Haram et al., "A Plasticene Lexicon," 110714.

15. Povinelli, *Geontologies.*

16. Povinelli, *Geontologies*, 37–38.

17. Davis, *Plastic Matter*, 53.

18. Synthetic frontiers expanding into an "empty" ocean fit with Povinelli's figure of the desert: "The space where life was, is not now, but could be if knowledges, techniques and resources where properly managed," Povinelli, *Geontologies,* 16.

19. Elizabeth A. Povinelli, *Economies of Abandonment: Social Belonging and Endurance in Late Liberalism* (Duke University Press, 2011), 25; Povinelli, *Geontologies.*

20. Povinelli, *Geontologies*, 5.

21. Povinelli, *Geontologies*, 174.

22. Marisol de la Cadena, *Earth Beings: Ecologies of Practice Across Andean Worlds* (Duke University Press, 2015); Povinelli, *Geontologies;* Zoe Todd, "Fish, Kin and Hope: Tending to Water Violations in *Amiskwaciwâskahikan* and Treaty Six Territory," *Afterall: A Journal of Art, Context and Enquiry* 43, (2017), 102–107.

23. Fujikane, *Mapping Abundance for a Planetary Future*, 22.

24. Povinelli, *Geontologies*, 20.

25. Indigenous justice is too often contained to being "brought in" on western terms. Even mountain ancestors given personhood status, for example, in New Zealand, are confined by settler state articulations of individual rights and determinations of where, exactly, a mountain begins and ends. Sammler, "Kauri and the Whale: Oceanic Matter and Meaning in New Zealand," 63–84.

26. Povinelli, *Geontologies*, 40–41. Here, plastic finds itself in the company of oil more broadly. For a meticulous discussion of oil's animacies complicating modern life/death and passive/ inert dualisms see Terra Schwerin Rowe, *Of Modern Extraction: Experiments in Critical Petrotheology* (London: Bloomsbury, 2022), especially chapter 4.

27. As meticulously detailed by Bernadette Bensaude-Vincent and Isabelle Stengers, the shift from analysis to synthesis in modern chemistry was caught up with the potential to upend the theory of vital force which held that organic substances could only be made by living bodies. The possibility of making life from nonlife, brought with it anxieties surrounding whether or not chemists had the power to make life, not whether their products would be antithetical to it. Bensaude-Vincent and Stengers, *A History of Chemistry*, 144–146.

28. David W. Laist, "Overview of the Biological Effects of Lost and Discarded Plastic Debris in the Marine Environment," *Marine Pollution Bulletin* 18, no. 6 (1987), 319–326.

29. Gregory, "Environmental Implications of Plastic Debris in Marine Settings—Entanglement, Ingestion, Smothering, Hangers-on, Hitch-Hiking and Alien Invasions," 2013–2025.

30. Emma M. Jepsen and P. J. Nico de Bruyn, "Pinniped Entanglement in Oceanic Plastic Pollution: A Global Review," *Marine Pollution Bulletin* 145 (2019), 295–305. In cataloguing oceanic forms of plastic harm, lines are cut not only between living and nonliving bodies, but between body insides and outsides. Even such formulations of entanglement are further separated from ingestion, where plastic enters digestive tracks. Laist, "Overview of the Biological Effects of Lost and Discarded Plastic Debris in the Marine Environment," 319–326. Just like the response to footage of Kamilo Beach after the Japan tsunami (chapter 1).

31. Jepsen and de Bruyn, "Pinniped Entanglement in Oceanic Plastic Pollution: A Global Review," 295–305.

32. Davis, *Plastic Matter*, 52. Plastic "disperses the inability to metabolize outward," as Davis continues, it "materializes the desire for *accumulation without metabolism*."

33. This poetic connection is epically elaborated in full by the graphic novel *The Rime of the Modern Mariner*, with the albatross-killing mariner stranded in the plastic seas of the Great Pacific Garbage Patch. Hayes, *The Rime of the Modern Mariner*.

34. A story that has come to dwell in my house as a children's book gifted to my daughter. Elisa Boxer, *One Turtle's Last Straw* (Crown Books for Young Readers, 2022).

35. Plastic Pollution Coalition, "Plastic Kills! Horror Short Film Contest," September 07, Available from https://www.plasticpollutioncoalition.org/blog/2023/9/7/plastic-kills-horror -short-film-contest-2023.

36. Anna Turns, "Saving the Albatross: 'The War Is Against Plastic and They Are Casualties on the Frontline,'" *The Guardian*, March 12, 2018.

37. Kimberly Y. Masibay, "Plastic or the Planet?" *Science World*, December 16, 2019.

38. For a prominent example, see the *National Geographic* multiyear web feature "Planet or Plastic?" *National Geographic* 2025.

39. Audra Mitchell and Aadita Chaudhury, "Worlding Beyond 'the' 'End' of 'the World': White Apocalyptic Visions and BIPOC Futurisms," *International Relations* 34, no. 3 (2020), 309–332; Masco, "The Crisis in Crisis," S65–76.

40. Claire Colebrook, *Who Would You Kill to Save the World?* (Lincoln: University of Nebraska Press, 2023).

41. The samples would then make one more Pacific crossing to California for storage in Algalita's archive.

42. Carson et al., "The Plastic-Associated Microorganisms of the North Pacific Gyre," 126–132.

43. This figure, however, is widely critiqued by oceanographers and marine ecologists who point out that plankton populations are incredibly variable, and that Algalita's method of weighing dehydrated plankton does not account for the large proportion of water in fully constituted plankton bodies and so on. Angelicque White, interview with the Author, 2012.

44. California Coastal Commission, "Prop 20 Turns 50! California Coastal Commission History," Available from https://www.coastal.ca.gov/history/.

45. Ocean Conservancy, "Charting a Course to Plastic Free Beaches: An Ocean Conservancy Policy Report Informed by 35 Years of International Coastal Cleanup Data," Ocean Conservancy, 2023.

46. *California Coastal Act of 1976.*

47. "California Coastal Cleanup Day History," California Coastal Commission, https://www.coastal.ca.gov/publiced/ccd/history.html#:~:text=The%20International%20Coastal%20Cleanup%20now,related%20to%20the%20marine%20environment.

48. Posters that include Yosemite Valley in 2017, and giant sequoia tree against background of Sierra Peaks in 2019. California Coastal Commission, "Poster Art," Available from https://www.coastal.ca.gov/publiced/ccd/poster_archive/.

49. Cig-egret might find synthetic kin in Pinar Yolda's stunningly provocative 2014 exhibit, *Ecosystem of Excess*, a project animated by Amarall-Zettler et al.'s plastisphere article. Illuminated test tubes imagine life forms evolving from the synthetic ooze of the Great Pacific Garbage Patch on a plastic planet not devoid of life, but decidedly devoid of humans. Pinar Yoldas, *Ecosystem of Excess*, 2014 (https://pinaryoldas.info/Ecosystem-of-Excess-2014).

50. A humbling reminder in just how anchored even new materialist posthuman agency is in bios/life capacities, here leveraged to reinforce divides where plastic is the opposite of life. Merely extending vitality to everything falls short of questioning the anchoring of vitality itself in western *bios*.

51. The animal agency of plastic is explored more elaborately in *The Majestic Plastic Bag*, a 2010 "mockumentary" produced by Los Angeles–based Heal the Bay, a nonprofit advocacy group dedicated to the protection of marine ecosystems. A clever parody of wildlife documentaries, the film traces the migration of a humble disposable white t-shirt bag, also known as the common grocery bag. Starting its life in a grocery store parking lot (store name obscured but

still obvious behind a carefully placed tree), the bag is born on the ground but soon takes flight. Narrowly escaping death in an urban park to join the inevitable path of water toward the sea, the bag displays the (un)expected characteristics of natural species as it flies, floats, and swims toward its destiny: the garbage patch in the middle of the Pacific Ocean. Delivered with the deadpan third-person narration of Jeremy Irons (the bag moves, but does not speak) and reinforced with a dramatic score, the film displays an impressive command of wildlife program tropes with Nature Channel accuracy. Heal the Bay reminds the public that "the plastic bag is not indigenous to the Pacific," again calling up the power of the invasive. In the end, the bag is as artificial as the landscapes it traverses: the asphalt jungle, manicured parks, and cement rivers of Los Angeles. The "thriving community" of plastic in the middle of the ocean is one that should not thrive. Jeremy Konner, *The Majestic Plastic Bag*, 2010.

52. Zettler et al., "Life in the "Plastisphere": Microbial Communities on Plastic Marine Debris," 7140.

53. Miriam C. Goldstein et al., "Increased Oceanic Microplastic Debris Enhances Oviposition in an Endemic Pelagic Insect," *Biology Letters* 8 (2012), 817–820.

54. Anna Turns, "The Photo That Made the Plastics Crisis Personal," *BBC*, June 2, 2023.

55. Liboiron, *Pollution Is Colonialism*, 104–106.

56. The turtle of straw-nose fame, too, is implicated in questionable relations. Reposted by popular YouTube channel Lad Bible, the video has racked up over forty-two million views (and links to a petition in conjunction with the Plastic Ocean foundation, for having the Great Pacific Garbage Patch "Trash Isle" recognized as a country by the United Nations). "Trash Isles: Turtle Gets Plastic Straw Removed from Its Nose by Rescuers | @LADbible," October 12, [cited 2024]. Available from https://youtu.be/vJmi_gwziy4?si=sTnxPM8NRW8gxcbv.

57. Povinelli, *Geontologies*, 11.

58. Davis, *Plastic Matter*, 7; Boetzkes, *Plastic Capitalism*.

59. Povinelli, Geontologies, 38. While Povinelli argues that well-rehearsed rejections of modern nature/culture divides too often leave intact metabolism-privileging constitutive difference between life and nonlife, there are scholars who have been doing just that: for example, Elizabeth Grosz's ontology of biology and the production of difference. Elizabeth Grosz, Becoming Undone: Darwinian Reflections on Life, Politics, and Art (Duke University Press, 2011).

60. For an astute analysis of the dynamics between the threads to and threats of the petrochemical industry, see Alice Mah, *Petrochemical Planet: Multiscalar Battles of Industrial Transformation* (Durham, NC: Duke, 2023).

61. Stephen Buranyi, "'We Are Just Getting Started': The Plastic-Eating Bacteria That Could Change the World," *The Guardian*, September 28, 2023.

62. Buranyi, "We Are Just Getting Started."

63. Yu Yang et al., "Comment on 'A Bacterium That Degrades and Assimilates Poly(ethylene terephthalate),'" *Science* 353, no. 6301 (2016), 759.

64. Shosuke Yoshida et al., "A Bacterium That Degrades and Assimilates Poly(ethylene terephthalate)," *Science* 351, no. 6278 (2016), 1196.

65. Yoshida et al., "A Bacterium That Degrades and Assimilates Poly(ethylene terephthalate)."

66. Zrimec et al., "Plastic-Degrading Potential Across the Global Microbiome Correlates with Recent Pollution Trends."

67. Sandy Ong, "The Living Things That Feast on Plastic," *ASBMB Today*, September 10, 2023.

68. Jennifer Gabrys, "Plastic and the Work of the Biodegradable," in *Accumulation: The Material Politics of Plastic*, ed. by Jennifer Gabrys, Gay Hawkins, Mike Michael (Routledge, 2013).

69. V. Tournier et al., "An Engineered PET Depolymerase to Break Down and Recycle Plastic Bottles," *Nature* 580, no. 7802 (2020), 216–219.

70. Ong, *The Living Things That Feast on Plastic*.

71. Many thanks to Samiee Espinoza for introducing me to *Love After the End* and drawing my attention specifically to the line in "History of the New World," where colonists are sent the telling message: "Your circle is not round." Adam Garnet Jones, "History of the New World," in *Love After the End: An Anthology of Two-Spirit and Indigiqueer Speculative Fiction*, ed. by Joshua Whitehead (Arsenal Pulp Press, 2020), 44.

72. Mah, "Future-Proofing Capitalism: The Paradox of the Circular Economy for Plastics," 121–142.

73. Chen, *Animacies*; Davis, *Plastic Matter*; Nicole Seymour, *Bad Environmentalism: Irony and Irreverence in the Ecological Age* (Minnesota University Press, 2018); Timothy Morton, "Guest Column: Queer Ecology," *PMLA/Publications of the Modern Language Association of America* 125, no. 2, 273–82. For Heather Davis, recognizing plastivorous bacteria as our "queer kin" holds the potential to upend the existing social and political order. Understood as relations of intensified responsibility that are irreducible to genetic lineage, kinship expands to new kinds of human descendants including plastic-eating microbial offspring. As Davis concludes, the challenge is to "find ways of embracing the inevitable emergence of multiple strange and beautiful life forms while holding chemical companies to account for the vast harms they are enacting on numerous bodies, human and nonhuman." Davis, *Plastic Matter*, 83.

74. Davis, *Plastic Matter*.

75. Davis, *Plastic Matter*; Todd, "Fish, Kin and Hope: Tending to Water Violations in *Amiskwaciwâskahikan* and Treaty Six Territory," 102–107.

76. Gabrys, "Plastic and the Work of the Biodegradable," 217.

LANDINGS

1. Anne C. Mulkern, "PR Battle over Plastic Bags Ends in Settlement," September 19, 2011.

2. Niamh Scallon. "'Frightening' Amount of Plastic Floating in North Pacific Gyre." *The Province*, July 2011, http://www.theprovince.com/technology/Frightening+amount+plastic+flo ating+North+Pacific+Gyre/5175183/story.html#ixzz1VZ03nQPo.

SYNTHETICS: PLACING TOGETHER/WITH

1.	Scallon, "'Frightening' Amount of Plastic Floating in North Pacific Gyre."

2.	Karen Barad, "Troubling Time/s and Ecologies of Nothingness: Re-Turning, Re-Membering, and Facing the Incalculable," *New Formations* 92 (2018), 59; Barad, "Diffracting Diffraction: Cutting Together-Apart," 168–187.

3.	In Michelle Huang's astute Barad-informed posthumanist reading of Ruth Ozeki's *A Tale for the Time Being* the garbage patch circulates as a transpacific Asian-American racial formation—violent Pacific histories discarded with plastic. Michelle N. Huang, "Ecologies of Entanglement in the Great Pacific Garbage Patch," *Journal of Asian American Studies* 20, no. 1 (2017), 95–117.

4.	Lived oceanic realities merge with academic conceptualizations of fluidities to push against linear narratives of western progress. In an explicit counterpoint to modern dialectics, Barbados-born poet Kamau Brathwaite offers "tidalectics" to describe ongoing relationalities rather than binary oppositions of land and sea, especially as they refuse to resolve into modern synthesis. As Elizabeth DeLoughrey writes, Brathwaite's "'tidal dialectic' resists the synthesizing telos of Hegel's dialectic by drawing from a cyclical model, invoking the continual movement and rhythm of the ocean. Tidalectics foreground 'alter/native' epistemologies to colonialism and capitalism, with their linear and materialist biases. In contradistinction to Western models of passive and empty space, such as terra (and aqua) nullius, which were used to justify territorial expansion, tidalectics reckons a space and time that requires an active and participatory engagement with the island seascape." Elizabeth DeLoughrey, "Revisiting Tidalectics: Irma, José, Maria," in *Tidalectics: Imagining an Oceanic Worldview Through Art and Science*, ed. by Stefanie Hessler (Cambridge, MA: The MIT Press, 2018). Tidalectics, then, are generative relations to place enacted as Brathwaite poetically performs the tensions of diasporic identities with the rhythms of the sea instead of those of the colonizer. Stefanie Hessler, *Tidalectics: Imagining an Oceanic Worldview through Art and Science* (Cambridge, MA; London: TBA21-Academy, 2018). If tidalectics are the "forward and back" of waves that challenge the linearity of modern processes even as they are made known through them, might the spiraling transformations of plastic in gyres also become refusals of the synthesis of modern chemistry? DeLoughrey, "Revisiting Tidalectics: Irma, José, Maria," in Helmreich, *A Book of Waves*; Elizabeth DeLoughrey, *Routes and Roots: Navigating Caribbean and Pacific Island Literatures* (Honolulu: University of Hawaii Press, 2007).

5.	In a noteworthy exception, a group of natural scientists led by analytic chemist Chelsea Rochman make a strong case for classifying plastic as hazardous waste because of its tendency to fragment into ever-smaller pieces that gather and leach chemical ingredients known to be toxic. Chelsea M. Rochman et al., "Classify Plastic Waste as Hazardous," *Nature* 494, no. 7436 (2013), 169–171.

6.	"The Ocean Cleanup LIVE from San Francisco Bay: Cleanup Operations Update," September 6, Available from https://www.youtube.com/watch?v=Cm1zIQhoa90.

7.	Alice Te Punga Somerville, writing on the nuances of what Indigeneity means in the Pacific, details forced compliance with colonial ontologies, epistemologies, and cartographies that ground sovereignty in claims to land.[1] Māori are forced by the narrow terrain of recognition to navigate tensions between continuity and diversity, where settler state demands of proving a long-standing fixed relation to land, seem to require disavowal of claims of being a traveling people with "the ability to adapt to a new landscape and also to remember previous homes." Alice Te Punga Somerville, *Once Were Pacific: Māori Connections to Oceania* (Minneapolis, MN: University of Minnesota Press, 2012), 86.

8. King, *The Black Shoals*, 3–4.

9. King, *The Black Shoals*, 56, 68.

10. King, *The Black Shoals*, 208.

11. Te Punga Somerville, *Once Were Pacific*, 141–142.

12. Diaz, "No Island Is an Island," 90–107.

13. Diaz, "No Island Is an Island," 102.

14. Diaz, "No Island Is an Island."

15. Diaz, "No Island Is an Island," 103.

16. It is not only oceanic peoples whose worlds have been "islanded," and who continue to occupy and exceed "cramped spaces." As Zoe Tood writes, "While Indigenous bodies and worlds are forced to occupy vanishingly small space in late liberalism and are erased nearly wholly in universalist understandings of contemporary phenomena like the Anthropocene, this cramped space is nonetheless simultaneously a space occupied by myriad, plural and pulsating cosmos, ontologies, and worlds." E. R. Johnson et al., "Geontographies: On Elizabeth Povinelli's Geontologies: A Requiem for Late Liberalism," *Environment and Planning C Politics and Space*, no. 8 (2019), 7.

17. What generative ways of thinking and being might emerge as radical water (Klaver, "Radical Water," 64–88; Melanie Yazzie & Cutcha Risling Baldy, "Introduction: Indigenous Peoples and the Politics of Water," *Decolonization: Indigeneity, Education & Society* 7, no. 1 (2018): 1–18) is further brought into conversation with anticolonial critiques of industrial chemicals (Liboiron, *Pollution Is Colonialism*; Murphy, "Alterlife and Decolonial Chemical Relations," 494–503; Isabelle Stengers, "Receiving the Gift: Earthly Events, Chemical Invariants, and Elemental Powers," in *Reactivating Elements: Chemistry, Ecology, Practice*, ed. by Dimitri Papadopoulos, Mara Puig de la Bellacasa, and Natasha Myers (Durham, NC: Duke University Press, 2022), 18–33).

18. See Christina Dunbar-Hester's *Oil Beach* on the ongoing transformations of San Pedro Bay as a node of local and global ecological violence. Christina Dunbar-Hester, *Oil Beach: How Toxic Infrastructure Threatens Life in the Ports of Los Angeles and Beyond* (Chicago; London: The University of Chicago Press, 2023).

19. LeMenager, *Living Oil*, 54–55.

20. Colin Ferris, "Fuels Paradise: THUMS Islands Help Big Oil Look Good," March 16 [cited 2024]. Available from https://webecoist.momtastic.com/2010/03/16/fuels-paradise-thums-islands-help-big-oil-look-good/.

21. Jonathan L. Clark, "Uncharismatic Invasives," *Environmental Humanities* 6, no. 1 (2015).

22. Dunbar-Hester, *Oil Beach*.

23. Editorial Team, "Sydney's Living Seawall: Giving Plastic a Purpose," *Carmen Busquets*, July 26, 2018.

24. Riley Taitingfong, "Editing Islands: (Re)Imagining Isolation in Gene Drive Science and Engagement," *Society for the Social Study of Science*, 2023.

25. Seung Hee Cho, "Island as Test Bed: Electric Futures in Jeju Island, South Korea," *Society for the Social Study of Science*, 2023.

26. Wong, *Refiguring the 'Renewable Energy Island' as Future Environment*.

27. George Hoberg, *The Resistance Dilemma: Place-Based Movements and the Climate Crisis* (Cambridge: MIT Press, 2021); Marco Grasso, *From Big Oil to Big Green: Holding the Oil Industry to Account for the Climate Crisis* (Cambridge: The MIT Press, 2022); Katayoun Shafiee, *Machineries of Oil: An Infrastructural History of BP in Iran* (Cambridge, MA: MIT Press, 2023).

28. Where Penelope Hardy's account of the nineteenth-century mappings of seafloors follows European impositions of terrestrial topographies on oceanic depths in the service of empire, dominant solutions to contemporary climate crises are rending such spaces newly extractable. Penelope K. Hardy, "Water as the Medium of Measurement: Mapping Global Oceans in the Nineteenth and Twentieth Centuries," in *Hydrohumanities: Water Discourse and Environmental Futures*, ed. by Kim De Wolff, Rina C. Faletti, and Ignacio López-Calvo (Berkeley, CA: University of California Press, 2022), 118–140. As Stefanie Hessler beautifully makes visible, "Modernist efforts of knowing are linked to extractive-destructive utility, from territorial expansionisms to industrial-scale exploitation of aquatic life." Stefanie Hessler, Armin Linke, and Bruno Latour, *Prospecting Ocean* (Cambridge, MA: The MIT Press, 2019), 24.

29. Lisa Han for example, traces the "extractive mediations" that "tame" the seafloor into a frontier for resource extraction to sustain green energy technologies. Lisa Han, *Deepwater Alchemy: Extractive Mediation and the Taming of the Sea Floor* (Minnesota University Press, 2024).

30. Bauer and Nielsen, "Oil Companies Are Ploughing Money into Fossil-Fuelled Plastics Production at a Record Rate—New Research."

31. Jason Ruiz, "A Hotel? A Bird Sanctuary? Future Uses of Long Beach's Oil Islands Being Discussed," *Long Beach Post*, September 16, 2023.

32. Jeremy Miller, "In Corpus Christi, Texas, Environmentalists Are Fighting a Slate of Proposed Desalination Plants," *Sierra*, June 13, 2023, accessed October 22, 2024.

33. Kamanamaikalani Beamer, "Opening Keynote Speaker," *Society for the Social Study of Science*, 2023.

Bibliography

ABC News. "Hidden, a 3.5 Million Trash Heap Lies in the Ocean." *ABC News*, August 6, 2008. https://abcnews.go.com/GMA/story?id=5524886&page=1.

ABC News. "Pacific Ocean to Receive Plastic Island." *ABC News*, June 30, 2010. https://abcnews.go.com/Technology/island-recycled-plastic-form-pacific/story?id=11054077.

Alaimo, Stacy. *Bodily Natures: Science, Environment, and the Material Self.* Bloomington: Indiana University Press, 2010.

Aluli Meyer, Manulani, Mehana Blaich Vaughan, and Malia Akutagawa. "Special Plenary: Aloha ʻĀina: Hawaiian Knowledge Today." Society for the Social Study of Science, 2023. https://www.youtube.com/watch?v=CXbsOEoHCTc&t=1410s.

Amaral-Zettler, Linda A. "Colonization of Plastic Marine Debris: The Known, the Unknown, and the Unknowable." In *Plastics and the Ocean: Origin, Characterization, Fate, and Impacts*, edited by Anthony L. Andrady. United States: John Wiley & Sons, 2022.

Amaral-Zettler, Linda A., Erik R. Zettler, and Tracy J. Mincer. "Ecology of the Plastisphere." *Nature Reviews Microbiology* 18, no. 3 (2020): 139–151.

Andrady, Anthony L. "Microplastics in the Marine Environment." *Marine Pollution Bulletin* 62, no. 8 (2011): 1596–1605.

Andrady, Anthony L. *Plastics and the Ocean: Origin, Characterization, Fate, and Impacts.* Hoboken: Wiley-Blackwell, 2022.

Armitage, David, and Alison Bashford. *Pacific Histories: Ocean, Land, People.* Basingstoke: Palgrave Macmillan, 2014.

Baldacchino, Godfrey, and Eric Clark. "Guest Editorial Introduction: Islanding Cultural Geographies." *Cultural Geographies* 20, no. 2 (2013): 129–134.

Barad, Karen. "Diffracting Diffraction: Cutting Together-Apart." *Parallax* 20, no. 3 (2014): 168–187.

Barad, Karen. *Meeting the Universe Halfway: Quantum Physics and the Entanglement of Matter and Meaning.* Durham, NC: Duke University Press, 2007.

Barad, Karen. "Posthumanist Performativity: Toward an Understanding of How Matter Comes to Matter." *Signs: Journal of Women in Culture and Society* 28, no. 3 (2003): 801–831.

Barad, Karen. "Troubling Time/s and Ecologies of Nothingness: Re-turning, Re-membering, and Facing the Incalculable." *new formations: a journal of culture/theory/politics*, 92 (2018): 56–86. https://muse.jhu.edu/article/689858.

Barney, Liz, and Michelle Broder Van Dyke. "Welcome to Hawaii's 'Plastic Beach.' One of the World's Dirtiest Places." *The Guardian*, January 10, 2020. https://www.theguardian.com/us-news/2020/jan/10/kamilo-beach-plastic-hawaii-pollution.

Barthes, Roland. *Mythologies*. New York: Hill and Wang, 1972.

Bauer, Fredric, and Tobias Dan Nielsen. "Oil Companies are Ploughing Money into Fossil-Fuelled Plastics Production at a Record Rate—New Research." *The Conversation*, November 2, 2021. https://theconversation.com/oil-companies-are-ploughing-money-into-fossil-fuelled-plastics-production-at-a-record-rate-new-research-169690.

Beamer, Kamanamaikalani. *No Mākou Ka Mana: Liberating the Nation*. Honolulu: Native Books, 2014.

Beamer, Kamanamaikalani. "Opening Keynote Speaker." Society for the Social Study of Science, November 9, 2023. https://www.youtube.com/watch?v=HweQRiVK7VQ.

Bennett, Jane. *Vibrant Matter: A Political Ecology of Things*. Durham, NC: Duke University Press, 2010.

Bensaude-Vincent, Bernadette, and Isabelle Stengers. *A History of Chemistry*. Cambridge: Harvard University Press, 1996.

Berton, Justin. "Continent-Size Toxic Stew of Plastic Trash Fouling Swath of Pacific Ocean." *SFGATE*, October 18, 2007. https://www.sfgate.com/green/article/Continent-size-toxic-stew-of-plastic-trash-2518237.php.

Boetzkes, Amanda. *Plastic Capitalism: Contemporary Art and the Drive to Waste*. Cambridge: The MIT Press, 2019.

Bowker, Geoffrey C., and Susan Leigh Star. *Sorting Things Out: Classification and its Consequences*. Cambridge: The MIT Press, 1999.

Boxer, Elisa. *One Turtle's Last Straw*. New York: Crown Books for Young Readers, 2022.

Boyer, Dominic. *Energopolitics: Wind and Power in the Anthropocene*. Durham, NC: Duke University Press, 2019.

Break Free From Plastic (BFFP). "About BFFP." Break Free From Plastic. Accessed February 24, 2023, https://www.breakfreefromplastic.org/about/.

Bruno Latour, Latour. *Pasteurization of France*. Cambridge: Harvard University Press, 1993.

Buranyi, Stephen. "The Plastic Backlash: What's Behind Our Sudden Rage—and Will It Make a Difference?" *The Guardian*, November 13, 2018. https://www.proquest.com/docview/2133066323?sourcetype=Newspapers.

Buranyi, Stephen. "'We Are Just Getting Started': The Plastic-Eating Bacteria That Could Change the World." *The Guardian*, September 28, 2023. https://www.theguardian.com/environment/2023/sep/28/plastic-eating-bacteria-enzyme-recycling-waste.

Burkhart, Brian. *Indigenizing Philosophy Through the Land: A Trickster Methodology for Decolonizing Environmental Ethics and Indigenous Futures*. East Lansing: Michigan State University Press, 2019.

Burnett, Christina Duffy. "The Edges of Empire and the Limits of Sovereignty: American Guano Islands." *American Quarterly* 57, no. 3 (2005): 779–803.

Byrd, Jodi A. *The Transit of Empire: Indigenous Critiques of Colonialism*. Minneapolis: University of Minnesota Press, 2011.

California Coastal Commission. "California Coastal Cleanup Day History." California Coastal Commission. Accessed May 26, 2024. Available from https://www.coastal.ca.gov/publiced/ccd/history.html#:~:text=The%20International%20Coastal%20Cleanup%20now,related%20to%20the%20marine%20environment.

California Coastal Commission. "Poster Art." California Coastal Commission. Accessed October 23, 2024. https://www.coastal.ca.gov/publiced/ccd/poster_archive/.

California Coastal Commission. "Prop 20 Turns 50! California Coastal Commission History." California Coastal Commission, accessed October 24, 2024. Available from https://www.coastal.ca.gov/history/.

Carson, Henry S., Magnus S. Nerheim, Katherine A. Carroll, and Marcus Eriksen. "The Plastic-Associated Microorganisms of the North Pacific Gyre." *Marine Pollution Bulletin* 75, no. 1–2 (2013): 126–132.

Chen, Mel Y. *Animacies: Biopolitics, Racial Mattering, and Queer Affect*. Durham, NC: Duke University Press, 2012.

Cho, Seung Hee. "Island as Test Bed: Electric Futures in Jeju Island, South Korea." Society for the Social Study of Science, December 2023.

Clark, Dan Elbert. "Manifest Destiny and the Pacific." *Pacific Historical Review* 1, no. 1 (1932): 1–17.

Clark, Jonathan L. "Uncharismatic Invasives." *Environmental Humanities* 6, no. 1 (2015): 29–52.

Clear Rivers. "CLEAR RIVERS | for a Plastic Free Sea!" Clear Rivers. Accessed April 20, 2024. Available from https://www.clearrivers.eu/.

Cohen, Lizabeth. *A Consumer's Republic: The Politics of Mass Consumption in Postwar America*. New York: Vintage, 2003.

Colebrook, Claire. *Who Would You Kill to Save the World?* Lincoln: University of Nebraska Press, 2023.

Coole, Diana H., and Samantha Frost, eds. *New Materialisms: Ontology, Agency, and Politics*. Durham, NC: Duke University Press, 2010.

Corcoran, Patricia L., Charles J. Moore, and Kelly Jazvac. "An Anthropogenic Marker Horizon in the Future Rock Record." *GSA Today* 24, no. 6 (2014): 4–8.

Cózar, Andrés, Fidel Echevarría, J. Ignacio González-Gordillo, et al. "Plastic Debris in the Open Ocean." *Proceedings of the National Academy of Sciences* 111, no. 28 (2014): 10239–10244.

Cronon, William. "The Trouble with Wilderness; Or, Getting Back to the Wrong Nature." *Environmental History* 1, no. 1 (1996): 7–28.

Crutzen, Paul J., and Eugene F. Stoermer. "The 'Anthropocene.'" *Global Change Newsletter* 41 (2000): 17–18.

da Cunha, Dilip. *The Invention of Rivers: Alexander's Eye and Ganga's Descent.* Philadelphia: University of Pennsylvania Press, 2019.

Daily Mail. "Rubbish Dump Found Floating in Pacific Ocean is Twice the Size of America." *Daily Mail*, February 6, 2008. Accessed October 24, 2024. https://www.dailymail.co.uk/news/article-512424/Rubbish-dump-floating-Pacific-Ocean-twice-size-America.html.

Davis, Heather. *Plastic Matter.* Durham, NC: Duke University Press, 2022.

Day, Robert H., and David G. Shaw. "Patterns in the Abundance of Pelagic Plastic and Tar in the North Pacific Ocean, 1976–1985." *Marine Pollution Bulletin* 18, no. 6 (1987): 311–316.

de la Cadena, Marisol. *Earth Beings: Ecologies of Practice Across Andean Worlds.* Durham, NC: Duke University Press, 2015.

De Wolff, Kim. "'Floating Things' and Methodological Drift: Accounting for Haunted Materialities in the North Pacific Ocean." *Social Studies of Science* 54, no. 4 (2024): 536–556.

De Wolff, Kim. "Plastic Naturecultures: Multispecies Ethnography and the Dangers of Separating Living from Nonliving Bodies." *Body & Society* 23, no. 3 (2017): 23–47.

De Wolff, Kim. "Plastivores and the Persistence of Synthetic Futures." In *Living in the Plastic Age: Perspectives from Humanities, Social Sciences and Natural Sciences*, edited by Johanna Kramm and Carolin Völker. Frankfurt: Campus, 2023.

De Wolff, Kim, Rina C. Faletti, and Ignacio López-Calvo, eds. *Hydrohumanities: Water Discourse and Environmental Futures.* Oakland: University of California Press, 2022.

Deleuze, Gilles, and Félix Guattari. *A Thousand Plateaus: Capitalism and Schizophrenia.* Minneapolis: Continuum, 2004.

DeLoughrey, Elizabeth. *Allegories of the Anthropocene.* Durham, NC: Duke University Press, 2019.

DeLoughrey, Elizabeth. "The Myth of Isolates: Ecosystem Ecologies in the Nuclear Pacific." *Cultural Geographies* 20, no. 2 (2013): 167–184.

DeLoughrey, Elizabeth. "Revisiting Tidalectics: Irma, José, Maria." In *Tidalectics: Imagining an Oceanic Worldview through Art and Science*, edited by Stefanie Hessler. Cambridge: The MIT Press, 2018.

DeLoughrey, Elizabeth. *Routes and Roots: Navigating Caribbean and Pacific Island Literatures.* Honolulu: University of Hawaii Press, 2007.

DeLoughrey, Elizabeth. "Submarine Futures of the Anthropocene." *Comparative Literature* 69, no. 1 (2017): 32–44.

DeLoughrey, Elizabeth. "Toward a Critical Ocean Studies for the Anthropocene." *English Language Notes* 57, no. 1 (2019): 21–36.

Diaz, Vincente. "No Island is an Island." In *Native Studies Keywords*, edited by Stephanie Nohelani Teves, Andrea Smith and Michelle Raheja, 90–107. Chicago: The University of Arizona Press, 2015.

DiGregorio, Barry E. "Tracking Plastic in the Oceans." *EARTH Magazine*, January 24, 2012. https://www.earthmagazine.org/article/tracking-plastic-oceans/.

Doyle, Julie. "Picturing the Clima(C)Tic: Greenpeace and the Representational Politics of Climate Change Communication." *Science as Culture* 16, no. 2 (2007): 129–150.

Dunbar-Hester, Christina. *Oil Beach: How Toxic Infrastructure Threatens Life in the Ports of Los Angeles and Beyond.* Chicago; London: The University of Chicago Press, 2023.

Ebbesmeyer, Curtis, and Eric Scigliano. *Flotsametrics and the Floating World.* New York: Harper-Collins Publishers, 2010.

Editorial Team. "Sydney's Living Seawall: Giving Plastic a Purpose." *Carmen Busquets*, July 26, 2018. https://www.carmenbusquets.com/journal/post/sydneys-living-seawall-giving-plastic-a -purpose#:~:text=So%20together%20with%20the%20Sydney,structure%20of%20native%20 mangrove%20trees.

Eriksen, Marcus, Laurent C. M. Lebreton, Henry S. Carson, et al. "Plastic Pollution in the World's Oceans: More Than 5 Trillion Plastic Pieces Weighing Over 250,000 Tons Afloat at Sea." *PLoS ONE* 9, no. 12 (2014).

Fache, Elodie, Pierre-Yves Le Meur, and Estienne Rodary. "Introduction: The New Scramble for the Pacific: A Frontier Approach." *Pacific Affairs* 94, no. 1 (2021): 57–76.

Farrelly, Trisia, Sy Taffel, and Ian Shaw, eds. *Plastic Legacies: Pollution, Persistence, and Politics.* Edmonton: Athabasca University Press, 2021.

Ferris, Colin. "Fuels Paradise: THUMS Islands Help Big Oil Look Good." *WebEcoist*, March 16, 2010. Accessed October 20, 2024. https://webecoist.momtastic.com/2010/03/16/fuels-paradise -thums-islands-help-big-oil-look-good/.

Finley, Carmel. *All the Fish in the Sea: Maximum Sustainable Yield and the Failure of Fisheries Management.* Chicago; London: The University of Chicago Press, 2011.

Finney, Carolyn. *Black Faces, White Spaces: Reimagining the Relationship of African Americans to the Great Outdoors.* Chapel Hill, NC: University of North Carolina Press, 2014.

Frederick, Eva. "Ninety-Nine Percent of Ocean Plastic Has Gone Missing." *Science*, January 3, 2020. Accessed October 24, 2024. https://www.science.org/content/article/ninety-nine-percent -ocean-plastic-has-gone-missing.

Freinkel, Susan. *Plastic: A Toxic Love Story.* Melbourne: Text Publishing, 2011.

Fujikane, Candace. *Mapping Abundance for a Planetary Future.* Durham, NC: Duke University Press, 2021.

Fuller, Sascha, Tina Ngata, Stephanie B. Borrelle, and Trisia Farrelly. "Plastics Pollution as Waste Colonialism in Te Moananui." *Journal of Political Ecology* 29, no. 1 (2022): 534–560.

Furuhata, Yuriko. "Of Dragons and Geoengineering: Rethinking Elemental Media." *Media + Environment* 1, no. 1 (2019). https://doi.org/10.1525/001c.10797.

Gabrys, Jennifer. "Plastic and the Work of the Biodegradable." In *Accumulation: The Material Politics of Plastic*, edited by Jennifer Gabrys, Gay Hawkins, and Mike Michael. London: Routledge, 2013.

Gabrys, Jennifer, Gay Hawkins, and Mike Michael, eds. *Accumulation: The Material Politics of Plastics*. London: Routledge, 2013.

Gardiner, Beth. "The Plastics Pipeline: A Surge of New Production is on the Way." *Yale Environment 360*, December 19, 2019. https://e360.yale.edu/features/the-plastics-pipeline-a-surge-of-new-production-is-on-the-way.

Garnet Jones, Adam. "History of the New World." In *Love After the End: An Anthology of Two-Spirit and Indigiqueer Speculative Fiction*, edited by Joshua Whitehead. Vancouver: Arsenal Pulp Press, 2020.

Geyer, Roland, Jenna R. Jambeck, and Kara Lavender Law. "Production, Use, and Fate of All Plastics Ever Made." *Science Advances* 3, no. 7 (2017): e1700782.

Ghosh, Amitav. *The Nutmeg's Curse: Parables for a Planet in Crisis*. Chicago: The University of Chicago Press, 2021.

Gibson, Eleanor. "The Ocean Cleanup Suspended as Device Breaks Down in Pacific Ocean." *Dezeen*, January 7, 2019. Accessed October 24, 2024. https://www.dezeen.com/2019/01/07/ocean-cleanup-suspended-pacific-plastic/.

Gillis, John R. "The Blue Humanities." *The Humanities* 34, no. 3 (2013). https://www.neh.gov/humanities/2013/mayjune/feature/the-blue-humanities.

Gillis, John R. *Islands of the Mind: How the Human Imagination Created the Atlantic World*. New York: Palgrave Macmillan, 2004.

Goldenberg, Maya J. *Vaccine Hesitancy: Public Trust, Expertise, and the War on Science*. Pittsburgh: University of Pittsburgh Press, 2021.

Goldstein, Miriam. "Does the 'Great Pacific Garbage Patch' Exist?" SEAPLEX, January 10, 2011. Archived January 14, 2011, at https://web.archive.org/web/20110201000000*/http://seaplexscience.com/2011/01/10/does-the-great-pacific-garbage-patch-exist/.

Goldstein, Miriam. "'Recycled Island' Not a Cure for Plastic Trash in Ocean." *Deep Sea News*, August 9, 2010. Accessed October 24, 2024. https://deepseanews.com/2010/08/recycled-island-not-cure-for-plastic-in-ocean/.

Goldstein, Miriam C., Henry S. Carson, and Marcus Eriksen. "Relationship of Diversity and Habitat Area in North Pacific Plastic-Associated Rafting Communities." *Marine Biology* 161, (2014): 1441–1453.

Goldstein, Miriam C., Marci Rosenberg, and Lanna Cheng. "Increased Oceanic Microplastic Debris Enhances Oviposition in an Endemic Pelagic Insect." *Biology Letters* 8, (2012): 817–820.

Grasso, Marco. *From Big Oil to Big Green: Holding the Oil Industry to Account for the Climate Crisis.* Cambridge: The MIT Press, 2022.

Gregory, Murray R. "Environmental Implications of Plastic Debris in Marine Settings— entanglement, Ingestion, Smothering, Hangers-on, Hitch-Hiking and Alien Invasions." *Philosophy Transactions of the Royal Society B* 364, no. 1526 (2009): 2013–2025.

Grosz, Elizabeth. *Becoming Undone: Darwinian Reflections on Life, Politics, and Art.* Durham, NC: Duke University Press, 2011.

Gunther, Marc. "In Defense of the Plastic Bag." *GreenBiz*, December 22, 2011. Accessed May 17, 2024. https://www.greenbiz.com/article/defense-plastic-bag.

Hamblin, Jacob Darwin. *Poison in the Well: Radioactive Waste in the Oceans at the Dawn of the Nuclear Age.* Piscataway: Rutgers University Press, 2008.

Hamouchene, Hamza and Katie Sandwell, eds. *Dismantling Green Colonialism: Energy and Climate Justice in the Arab Region.* London: Pluto Press, 2023.

Han, Lisa. *Deepwater Alchemy: Extractive Mediation and the Taming of the Sea Floor.* Minneapolis: Minnesota University Press, 2024.

Haram, Linsey E., James T. Carlton, Gregory M. Ruiz, and Nikolai A. Maximenko. "A Plasticene Lexicon." *Marine Pollution Bulletin* 150, (2020): 110714.

Haraway, Donna. "A Manifesto for Cyborgs: Science, Technology, and Socialist Feminism in the 1980s." In *The Haraway Reader*, 7–45. New York: Routledge, 2004.

Haraway, Donna. "Situated Knowledges: The Science Question in Feminism and the Privilege of Partial Perspective." *Feminist Studies* 14, no. 3 (1988): 575–599.

Haraway, Donna. *When Species Meet.* Minneapolis: University of Minnesota Press, 2008.

Hardy, Penelope K. "Water as the Medium of Measurement: Mapping Global Oceans in the Nineteenth and Twentieth Centuries." In *Hydrohumanities: Water Discourse and Environmental Futures*, edited by Kim De Wolff, Rina C. Faletti and Ignacio López-Calvo, 118–140. Berkeley: University of California Press, 2022.

Harris, Joe, and Martín Morazzo. *Great Pacific Volume 1: Trashed!* Image Comics, 2013a.

Harris, Joe, and Martín Morazzo. *Great Pacific Volume 2: Nation Building.* Image Comics, 2013b.

Harris, Joe, and Martín Morazzo. *Great Pacific Volume 3: Big Game Hunters.* Image Comics, 2015.

Hau'ofa, Epeli. "Our Sea of Islands." *The Contemporary Pacific* 6, no. 1 (1994): 147–161.

Hawaii Statewide GIS Program. "Ahupuaa." (2024). https://geoportal.hawaii.gov/datasets/07624 815fc7d42d4b23c527d20ad2f58_1/explore?location=18.960920%2C-155.676299%2C12.17.

Hawkins, Gay. *The Ethics of Waste: How We Relate to Rubbish.* Lanham: Rowman & Littlefield, 2006.

179   *Bibliography*

Hawkins, Gay. "The Politics of Bottled Water." *Journal of Cultural Economy* 2, no. 1–2 (2009): 183–195.

Hayes, Nick. *The Rime of the Modern Mariner*. New York: Viking, 2012.

Helmreich, Stefan. *Alien Ocean: Anthropological Voyages in Microbial Seas*. Berkeley, CA: University of California Press, 2009.

Helmreich, Stefan. "An Anthropologist Underwater: Immersive Soundscapes, Submarine Cyborgs, and Transductive Ethnography." *American Ethnologist* 34, no. 4 (2007): 621–641.

Helmreich, Stefan. *A Book of Waves*. Durham, NC: Duke University Press, 2023.

Helmreich, Stefan. "Nature/Culture/Seawater." *American Anthropologist* 113, no. 1 (2011): 132–144.

Hessler, Stefanie. *Tidalectics: Imagining an Oceanic Worldview Through Art and Science*. Cambridge: TBA21-Academy, 2018.

Hessler, Stefanie, Armin Linke, and Bruno Latour. *Prospecting Ocean*. Cambridge: The MIT Press, 2019.

Hoberg, George. *The Resistance Dilemma: Place-Based Movements and the Climate Crisis*. Cambridge: The MIT Press, 2021.

Hohn, Donovan. *Moby-Duck: The True Story of 28,800 Bath Toys Lost at Sea and of the Beachcombers, Oceanographers, Environmentalists, and Fools, Including the Author, Who Went in Search of Them*. New York: Viking, 2011.

Hohn, Donovan. "Sea of Trash." *The New York Times*, June 22, 2008. Accessed May 17, 2024. https://www.nytimes.com/2008/06/22/magazine/22Plastics-t.html.

Hollin, Gregory, Isla Forsyth, Eva Giraud, and Tracey Potts. "(Dis)Entangling Barad: Materialisms and Ethics." *Social Studies of Science* 47, no. 6 (2017): 918–941.

Hoshaw, Lindsey. "Afloat in the Ocean, Expanding Islands of Trash." *The New York Times*, November 9, 2009. Accessed May 17, 2024. https://www.nytimes.com/2009/11/10/science/10patch.html.

Howe, Cymene. *Ecologics: Wind and Power in the Anthropocene*. Durham, NC: Duke University Press, 2019.

Huang, Michelle N. "Ecologies of Entanglement in the Great Pacific Garbage Patch." *Journal of Asian American Studies* 20, no. 1 (2017): 95–117.

Huggett, R. J. "Ecosphere, Biosphere, Or Gaia? What to Call the Global Ecosystem." *Global Ecology and Biogeography* 8, (2001): 425–431.

iADYS. "Jellyfishbot a Solution for Water Cleaning, Depollution, & Preservation." iADYS. Accessed May 17, 2024. https://www.iadys.com/jellyfishbot/.

Ingersoll, Karin Amimoto. *Waves of Knowing: A Seascape Epistemology*. Durham, NC: Duke University Press, 2016.

Jackson, Zakiyyah Iman. *Becoming Human: Matter and Meaning in an Antiblack World.* New York: New York University Press, 2020.

Jackson, Zakiyyah Iman. "Outer Worlds: The Persistence of Race in Movement 'Beyond the Human.'" *GLQ* 21, no. 2 (2015): 215–218.

Jambeck, Jenna R., Roland Geyer, Chris Wilcox, et al. "Plastic Waste Inputs from Land into the Ocean." *Science* 347, no. 6223 (2015): 768–771.

James, Jennifer C. "'Buried in Guano': Race, Labor, and Sustainability." *American Literary History* 24, no. 1 (2012): 115–142.

Jepsen, Emma M., and P. J. Nico de Bruyn. "Pinniped Entanglement in Oceanic Plastic Pollution: A Global Review." *Marine Pollution Bulletin* 145, (2019): 295–305.

Johnson, E. R., G. Kindervater, Z. Todd, K. Yusoff, K. Woodward, and E. A. Povinelli. "Geontographies: On Elizabeth Povinelli's Geontologies: A Requiem for Late Liberalism." *Environment and Planning C Politics and Space* 8 (2019): 1319–1342.

Jordan, Chris. *Albatross.* 2017. https://www.albatrossthefilm.com/watch-albatross.

Jue, Melody. "Floating Architectures." In *The Routledge Companion to Media and Risk*, edited by Bishnupriya Ghosh and Bhaskar Sarkar, 315–327. Routledge, 2020a.

Jue, Melody. *Wild Blue Media: Thinking Through Seawater.* Durham, NC: University Press, 2020b.

Jue, Melody, and Rafico Ruiz, eds. *Saturation: An Elemental Politics.* Durham, NC: Duke University Press, 2021.

Kennedy, John F. "The New Frontier." *Democratic National Convention.* July 15, 1960. https://www.presidency.ucsb.edu/documents/address-accepting-the-democratic-nomination-for-president-the-memorial-coliseum-los.

Kimmerer, Robin Wall. *Braiding Sweetgrass: Indigenous Wisdom, Scientific Knowledge and the Teachings of Plants.* London: Penguin Books, 2020.

King, Tiffany Lethabo. *The Black Shoals: Offshore Formations of Black and Native Studies.* Durham, NC: Duke University Press, 2019.

Klaver, Irene. "Radical Water." In *Hydrohumanities: Water Discourse and Environmental Futures*, edited by Kim De Wolff, Rina C. Faletti and Ignacio López-Calvo, 64–88. Oakland, CA: University of California Press, 2022.

Klaver, Irene J. "Meander(ing) Multiplicity." In *Water Scarcity, Security and Democracy: A Mediterranean Mosaic*, edited by Francesca de Châtel, Gail Holst-Warhaft and Tammo Steenhuis, 38–47. Global Water Partnership Mediterranean, Cornell University, and Atkinson Center for a Sustainable Future, 2014.

Knoester, Ramon. "Recycled Living Spaces." *The Journal of Ocean Technology* 11, no. 2 (2016): 52–58.

Konner, Jeremy. *The Majestic Plastic Bag.* Heal the Bay. 2010. https://www.youtube.com/watch?v=GLgh9h2ePYw.

Kostigen, Thomas M. "The World's Largest Dump: The Great Pacific Garbage Patch." *Discover Magazine*, July 9, 2008. Accessed October 24, 2024. https://www.discovermagazine.com/environment/the-worlds-largest-dump-the-great-pacific-garbage-patch.

Kroll, Gary. *America's Ocean Wilderness: A Cultural History of Twentieth-Century Exploration*. La Vergne: University Press of Kansas, 2008.

LAD Bible TV. "Trash Isles: Turtle Gets Plastic Straw Removed from Its Nose by Rescuers." *YouTube*, October 12, 2017. Accessed October 24, 2024. https://youtu.be/vJmi_gwziy4?si=sTnxPM8NRW8gxcbv.

Laist, David W. "Overview of the Biological Effects of Lost and Discarded Plastic Debris in the Marine Environment." *Marine Pollution Bulletin* 18, no. 6 (1987): 319–326.

Latour, Bruno. *Reassembling the Social: An Introduction to Actor-Network-Theory*. Oxford, UK: Oxford University Press, 2005.

Latour, Bruno. *We Have Never Been Modern*. New York: Harvester Wheatsheaf, 1993.

Latour, Bruno. *What is the Style of Matters of Concern? Two Lectures in Empirical Philosophy*. Assen: Koninklijke Van Gorcum, 2008.

Latour, Bruno. "Why Has Critique Run Out of Steam? From Matters of Fact to Matters of Concern." *Critical Inquiry* 30, no. 2 (2004): 225–248.

Lear, G., J. M. Kingsbury, S. Franchini, et al. "Plastics and the Microbiome: Impacts and Solutions." *Environmental Microbiome* 16, no. 2 (2021). https://doi.org/10.1186/s40793-020-00371-w.

Lebreton, Laurent, Boyen Slat, Francesco. Ferrari, et al. "Evidence That the Great Pacific Garbage Patch Is Rapidly Accumulating Plastic." *Scientific Reports* 8, no. 1 (2018): 4666–15. https://doi.org/10.1038/s41598-018-22939-w.

LeMenager, Stephanie. *Living Oil: Petroleum Culture in the American Century*. Oxford: Oxford University Press, 2014.

Lerner, Sharon. "Waste Only: How the Plastics Industry Is Fighting to Keep Polluting the World." *The Intercept*, 2019. https://theintercept.com/2019/07/20/plastics-industry-plastic-recycling/.

Leslie, Esther. *Synthetic Worlds: Nature, Art and the Chemical Industry*. London: Reaktion Books, 2005.

Lewis, Justin Matthew Wren, Andy Williams, and Robert Arthur Franklin. "A Compromised Fourth Estate? UK News Journalism, Public Relations and News Sources." *Journalism Studies* 9, no. 1 (2008): 1–20.

Liboiron, Max. "How the Ocean Cleanup Array Fundamentally Misunderstands Marine Plastics and Causes Harm." Discard Studies, June 5, 2015. Accessed October 24, 2024. https://discardstudies.com/2015/06/05/how-the-ocean-clean-up-array-fundamentally-misunderstands-marine-plastics-and-causes-harm/.

Liboiron, Max. *Pollution Is Colonialism*. Durham, NC: Duke University Press, 2021.

Liboiron, Max, and Josh Lepawsky. *Discard Studies: Wasting, Systems, and Power*. Cambridge: The MIT Press, 2022.

Lincoln, Noa K., Mehana Blaich Vaughan, and Natalie Kurashima. "Hawai'i." In *Islands & Cultures*, edited by Kamanamaikalani Beamer, Te Maire Tau and Peter Morrison Vitousek, 35–75. New Haven; London: Yale University Press, 2022.

Linton, Jamie. *What Is Water?: The History of a Modern Abstraction*. Vancouver: UBC Press, 2010.

Lorde, Audre. "The Master's Tools Will Never Dismantle the Master's House." In *This Bridge Called My Back, Fourth Edition*, edited by Cherríe Moraga and Gloria Anzaldúa, 94–97. SUNY Press, 2021.

Luanaigh, Cian. "Recycled Island: Plastic Fantastic?" *The Guardian*, August 7, 2010. https://www.theguardian.com/environment/2010/aug/08/recycled-island-plastic-waste-pacific.

MacBride, Samantha. *Recycling Reconsidered: The Present Failure and Future Promise of Environmental Action in the United States*. Cambridge: The MIT Press, 2011.

Mah, Alice. "Future-Proofing Capitalism: The Paradox of the Circular Economy for Plastics." *Global Environmental Politics* 21, no. 2 (2021): 121–142.

Mah, Alice. *Petrochemical Planet: Multiscalar Battles of Industrial Transformation*. Durham, NC: Duke, 2023.

Mah, Alice. *Plastic Unlimited: How Corporations Are Fueling the Ecological Crisis and What We Can Do About It*. New York: Polity Press, 2022.

Masco, Joseph. "The Artificial World." In *Reactivating Elements: Chemistry, Ecology, Practice*, edited by Dimitris Papadopoulos, María Puig de la Bellacasa, and Natasha Myers, 131–150. Durham, NC: Duke University Press, 2022.

Masco, Joseph. "The Crisis in Crisis." *Current Anthropology* 58, no. S15 (2017): S65–76.

Masibay, Kimberly Y. "Plastic or the Planet?" *Science World*, December 16, 2019. https://scienceworld.scholastic.com/issues/2019-20/121619/plastic-or-the-planet.html?language=english#1000L.

Matsuda, Matt K. "*AHR Forum*: The Pacific." *American Historical Review* 111, (2006): 758–777.

Mawyer, Alexander. "Floating Islands, Frontiers, and Other Boundary Objects on the Edge of Oceania's Futurity." *Pacific Affairs* 94, no. 1 (2021): 123–144.

Meikle, Jeffrey L. *American Plastic: A Cultural History*. New Brunswick, NJ: Rutgers University Press, 1995.

Mentz, Steve. "Blue Humanities." In *Posthuman Glossary*, edited by Rosi Braidotti and Maria Hlavajova, 69–72. London: Bloomsbury, 2018.

Messenger, Stephen. "'Recycled Island' Turns Ocean Plastic into a Paradise." *Treehugger*, June 28 2010. Accessed October 24, 2024. https://www.treehugger.com/recycled-island-turns-ocean-plastic-into-a-paradise-4858242.

Miller, Jeremy. "In Corpus Christi, Texas, Environmentalists Are Fighting a Slate of Proposed Desalination Plants." *Sierra*, June 13, 2023. Accessed October 24, 2024. https://www.sierraclub.org/sierra/2023-2-summer/feature/corpus-christi-texas-environmentalists-are-fighting-desalination.

Mills, Eric L. *The Fluid Envelope of Our Planet: How the Study of Ocean Currents Became a Science.* Toronto: University of Toronto Press, 2009.

Mitchell, Audra, and Aadita Chaudhury. "Worlding Beyond 'the' 'end' of 'the World': White Apocalyptic Visions and BIPOC Futurisms." *International Relations* 34, no. 3 (2020): 309–332.

Moore, Charles J. "Choking the Oceans with Plastic." *The New York Times*, August 25, 2014. Accessed October 24, 2014. https://www.nytimes.com/2014/08/26/opinion/choking-the-oceans-with-plastic.html.

Moore, Charles, and Cassandra Phillips. *Plastic Ocean: How a Sea Captain's Chance Discovery Launched a Determined Quest to Save the Oceans.* New York: Avery, 2011.

Morton, Timothy. "Guest Column: Queer Ecology." *PMLA/Publications of the Modern Language Association of America* 125, no. 2 (2010): 273–82.

Mueller, Gustav E. "The Hegel Legend of Thesia-Antithesis-Synthesis." *Journal of the History of Ideas* 19, no. 1/4 (1958): 411–414.

Mukerji, Chandra. *Territorial Ambitions and the Gardens of Versailles.* Cambridge: Cambridge University Press, 1997.

Mukerji, Chandra. "The Territorial State as a Figured World of Power: Strategics, Logistics, and Impersonal Rule." *Sociological Theory* 28 no. 4 (2010): 402–424.

Mulkern, Anne C. "PR Battle over Plastic Bags Ends in Settlement." *New York Times.* September 19, 2011. Accessed October 25, 2024. https://archive.nytimes.com/www.nytimes.com/gwire/2011/09/19/19greenwire-pr-battle-over-plastic-bags-ends-in-court-sett-61498.html?pagewanted=2#:~:text=A%20lawsuit%20over%20criticisms%20of,manufacturer%20ChicoBag%20of%20Chico%2C%20Calif.

Murcott, Toby. "Science Journalism: Toppling the Priesthood." *Nature* 459, (2009): 1054–1055.

Murphy, Michelle. "Alterlife and Decolonial Chemical Relations." *Cultural Anthropology* 32, no. 4 (2017): 494–503.

Murphy, Michelle. "Studying Unformed Objects: Deviation." Member Voices, *Fieldsights*, July 15, 2013. https://culanth.org/fieldsights/studying-unformed-objects-deviation.

Myers, Lynne. "The 8th Continent by Lenka Petráková is a Floating Station Concept to Clean Up our Oceans." *Designboom*, January 6, 2021. Accessed October 24, 2024. https://www.designboom.com/architecture/the-8th-continent-lenka-petrakova-clean-up-oceans-01-06-2020/.

National Geographic. "Planet or Plastic?" Accessed April 14, 2025. https://www.nationalgeographic.com/environment/topic/planetorplastic.

National Oceanic and Atmospheric Administration (NOAA). "Debunking Myths About Garbage Patches." National Oceanic and Atmospheric Administration, April 11, 2024. Accessed

October 18, 2024. https://response.restoration.noaa.gov/about/media/debunking-myths-about
-garbage-patches.html.

National Oceanic and Atmospheric Administration (NOAA). "What Are Marine Microbes?"
National Oceanic and Atmospheric Administration. Accessed October 19, 2024. https://
oceanexplorer.noaa.gov/facts/marinemicrobes.html.

Neimanis, Astrida. *Bodies of Water: Posthuman Feminist Phenomenology.* London: Bloomsbury
Publishing Plc, 2017.

Nelson, Bryan. "Hawaii-Sized Recycled Island to Be Built from Ocean Garbage Patch."
Mother Nature Network, April 1, 2011. Archived April 4, 2011 at https://web.archive.org/web
/20111113211723/http://www.mnn.com/earth-matters/wilderness-resources/stories/hawaii-sized
-recycled-island-to-be-built-from-ocean-garba.

Newitz, Annalee. "Lies You've Been Told About the Pacific Garbage Patch." *Gizmodo* May 21,
2012. Accessed October 24, 2024. https://gizmodo.com/lies-youve-been-told-about-the-pacific
-garbage-patch-5911969.

Ocean Conservancy. "Charting a Course to Plastic Free Beaches: An Ocean Conservancy Policy
Report Informed by 35 Years of International Coastal Cleanup Data." Ocean Conservancy, 2023.
https://oceanconservancy.org/trash-free-seas/international-coastal-cleanup/plastic-free-beaches/.

Olson, Steve. *Evolution in Hawaii: A Supplement to Teaching About Evolution and the Nature of
Science.* Washington DC: National Academies Press, 2004.

Ong, Sandy. "The Living Things That Feast on Plastic." *ASBMB Today*, September 3, 2023.
Accessed October 24, 2024. https://www.asbmb.org/asbmb-today/science/091023/the-living
-things-that-feast-on-plastic.

Oreskes, Naomi, and Erik M. Conway. *Merchants of Doubt: How a Handful of Scientists Obscured
the Truth on Issues from Tobacco Smoke to Global Warming.* New York: Bloomsbury Press, 2010.

Pangaea Exploration. "Sea Dragon." Pangaea Exploration. Accessed April 9, 2025. https://www
.panexplore.com/about-us/sea-dragon/.

Papadopoulos, Dimitris, María Puig de la Bellacasa, and Natasha Myers. *Reactivating Elements:
Chemistry, Ecology, Practice.* Durham, NC: Duke University Press, 2022.

Petráková, Lenka. "The 8th Continent." *LenkaPetrakova.com.* Accessed December 3, 2023. https://
lenkapetrakova.com/the-8th-continent-info.

Plastic Pollution Coalition. "Plastic Kills! Horror Short Film Contest." Plastic Pollution Coalition,
September 07, 2023. https://www.plasticpollutioncoalition.org/blog/2023/9/7/plastic-kills-horror
-short-film-contest-2023.

Povinelli, Elizabeth A. *Economies of Abandonment: Social Belonging and Endurance in Late Liberalism.* Durham, NC: Duke University Press, 2011.

Povinelli, Elizabeth A. *Geontologies: A Requiem to Late Liberalism.* Durham, NC: Duke University
Press, 2016.

Powell, Miri. "Synthetics of Empire: Plastics, Toxicity and Waste in the Pacific." Society for the Social Study of Science, 2022.

Pravda. "'Trash Island' Discovered in the Pacific Ocean." *English Pravda Online*, February 24, 2004. Archived April 14, 2004, at https://web.archive.org/web/20070105205051/http://newsfromrussia.com/science/2004/02/24/52469.html.

Puig de la Bellacasa, Maria. *Matters of Care: Speculative Ethics in More than Human Worlds*. Minneapolis, MN: University Of Minnesota Press, 2017.

Rabesandratana, Tania. "Report Traces Surge in Ocean Plastic Studies." *Science* 372, no. 6548 (2021): 1249.

Reno, Joshua. *Military Waste: The Unexpected Consequences of Permanent War Readiness*. Oakland: University of California Press, 2019.

Rillig, Matthias C., Kim Shin Woong, and Yong-Guan Zhu. "The Soil Plastisphere." *Nature Reviews Microbiology* 22, (2024): 64–74.

Roberts, Brian Russell. *Borderwaters: Amid the Archipelagic States of America*. Durham, NC: Duke University Press, 2021.

Robyn J. Wright, Gabriel Erni-Cassola, Vinko Zadjelovic, et al. "Marine Plastic Debris: A New Surface for Microbial Colonization." *Environmental Science & Technology* 54, no. 19 (2020): 11657–11672.

Rochman, Chelsea M., Mark Anthony Browne, Benjamin S. Halpern, et al. "Classify Plastic Waste as Hazardous." *Nature* 494, no. 7436 (2013): 169–171.

Roosth, Sophia. *Synthetic: How Life Got Made*. Chicago: University of Chicago Press, 2017.

Rowe, Terra Schwerin. *Of Modern Extraction: Experiments in Critical Petro-theology*. London: Bloomsbury, 2022.

Rozwadowski, Helen M. "Arthur C. Clarke and the Limitations of the Ocean as a Frontier." *Environmental History* 17, no. 3 (2012): 578–602.

Rozwadowski, Helen M. *Vast Expanses: A History of the Oceans*. London: Reaktion Books, 2018.

Ruiz, Jason. "A Hotel? A Bird Sanctuary? Future Uses of Long Beach's Oil Islands being Discussed." *Long Beach Post*, September 16, 2023. Accessed October 24, 2024. https://lbpost.com/news/oil-island-long-beach-reused-discussed/.

Russo, Daniella. "Plastic Pollution, Not Marine Debris!" Plastic Pollution Coalition, May 27, 2011a. Archived April 30, 2011 at https://web.archive.org/web/20110430200607/http://plasticpollutioncoalition.org/2011/03/plastic-pollution-not-marine-debris/.

Russo, Daniella. "Wrapping Up 5IMDC, Honolulu, Hawaii." Plastic Pollution Coalition, March 29, 2011b. Archived June 4, 2011 at https://web.archive.org/web/20110604060522/http://plasticpollutioncoalition.org/2011/03/wrapping-up-5imdc-honolulu-hawaii/.

Salesa, Damon. "The Pacific in Indigenous Time." In *Pacific Histories: Ocean, Land, People*, edited by David Armitage and Alison Bashford, 31–52. Palgrave Macmillan, 2014.

Sammler, Katherine. "Kauri and the Whale: Oceanic Matter and Meaning in New Zealand." In *Blue Legalities: The Life and Laws of the Sea*, edited by Irus Braverman and Elizabeth R. Johnson, 63–84. Durham, NC: Duke University Press, 2020.

Scallon, Niamh. "'Frightening' Amount of Plastic Floating in North Pacific Gyre." *The Province*, July 2011.

Schwartz, Ariel. "Electrolux Turning Plastic from the Ocean into Vacuum Cleaners." *Fast Company*, 2010. https://www.fastcompany.com/1663579/electrolux-turning-plastic-ocean-vacuum-cleaners.

Seton, Maria, Joanne Whittaker, and Simon Williams. "What Are Lost Continents, and Why Are We Discovering So Many?" *The Conversation*, November 24, 2019. Accessed October 24, 2024. https://theconversation.com/what-are-lost-continents-and-why-are-we-discovering-so-many-126355.

Seymour, Nicole. *Bad Environmentalism: Irony and Irreverence in the Ecological Age*. Minneapolis, MN: Minnesota University Press, 2018.

Shafiee, Katayoun. *Machineries of Oil: An Infrastructural History of BP in Iran*. Cambridge, MA: The MIT Press, 2023.

Shih, Ashanti. "The Most Perfect Natural Laboratory in the World: Making and Knowing Hawaii National Park—Ashanti Shih, 2019." *History of Science* 57, no. 4 (2019): 493–517.

Shotwell, Alexis. *Against Purity: Living Ethically in Compromised Times*. Minneapolis, MN; London: University of Minnesota Press, 2016.

Sicotte, Diane M. "From Cheap Ethane to a Plastic Planet: Regulating an Industrial Global Production Network." *Energy Research & Social Science* 66, (2020): 101479.

Sivasundaram, Sujit. "Science." In *Pacific Histories: Ocean, Land, People*, edited by David Armitage and Alison Bashford, 237–261. Palgrave Macmillan, 2014.

Slat, Boyan. "How the Oceans Can Clean Themselves." YouTube, October 24, 2012. Accessed November 10, 2023. https://www.youtube.com/watch?v=ROW9F-c0kIQ&t=137s.

Starosielski, Nicole. "The Elements of Media Studies." *Media + Environment* 1, no. 1 (2019). https://doi.org/10.1525/001c.10780.

Steinberg, Philip. *The Social Construction of the Ocean*. Cambridge, UK: Cambridge University Press, 2001.

Steinberg, Philip, and Kimberley Peters. "Wet Ontologies, Fluid Spaces: Giving Depth to Volume Through Oceanic Thinking." *Environment and Planning. D: Society & Space* 33, no. 2 (2015): 247–264.

Stengers, Isabelle. "Receiving the Gift: Earthly Events, Chemical Invariants, and Elemental Powers." In *Reactivating Elements: Chemistry, Ecology, Practice*, edited by Dimitri Papadopoulos, Mara Puig de la Bellacasa, and Natasha Myers, 18–33. Durham, NC: Duke University Press, 2022.

Strasser, Susan. *Waste and Want: A Social History of Trash*. New York: Metropolitan Books, 1999.

Tabuchi, Hiroko. "How a New Treaty Could Clean Up Plastic Waste." *The New York Times*, March 2, 2022. Accessed February 24, 2023. https://www.nytimes.com/2022/03/02/climate /global-plastics-recycling-treaty.html.

Taffel, Sy. "Communicative Capitalism, Technological Solutionism, and the Ocean Cleanup." In *Plastic Legacies: Pollution, Persistence, Politics*, edited by Trisia Farrelly, Sy Taffel and Ian Shaw. Athabasca, AB: Athabasca University Press, 2022.

Taitingfong, Riley. "Editing Islands: (Re)Imagining Isolation in Gene Drive Science and Engagement." Society for the Social Study of Science, 2023.

Te Punga Somerville, Alice. *Once Were Pacific: Māori Connections to Oceania*. Minneapolis, MN: University of Minnesota Press, 2012.

Teaiwa, Teresia. "To Island." In *A World of Islands: An Island Studies Reader*, edited by Godfrey Baldacchino, 514. Charlottetown, PEI: University of Prince Edward Island and Agenda Academic, 2007.

The Bryant Park Project. "Garbage Mass Is Growing in the Pacific." *National Public Radio*, March 26, 2008. https://www.npr.org/2008/03/26/89099470/garbage-mass-is-growing-in-the-pacific.

The Ocean Cleanup. "Oceans • the Ocean Cleanup." The Ocean Cleanup. Accessed October 24, 2024. https://theoceancleanup.com/oceans/.

The Ocean Cleanup. "The Ocean Cleanup LIVE from San Francisco Bay: Cleanup Operations Update." YouTube, September 6, 2024. Accessed October 24, 2024. https://www.youtube.com /watch?v=Cm1zIQhoa90.

The Ocean Cleanup. *2022 Annual Report*. The Ocean Cleanup. Accessed December 3, 2023. https://assets.theoceancleanup.com/app/uploads/2023/06/TheOceanCleanup_AnnualReport _2022.pdf.

The Ocean Cleanup. "The Great Pacific Garbage Patch." The Ocean Cleanup. Accessed October 18, 2024. https://theoceancleanup.com/great-pacific-garbage-patch/.

The Seasteading Institute. "About—The Seasteading Institute." The Seasteading Institute. Accessed May 17, 2024. https://www.seasteading.org/about/.

Todd, Zoe. "Fish, Kin and Hope: Tending to Water Violations in *Amiskwaciwâskahikan* and Treaty Six Territory." *Afterall: A Journal of Art, Context and Enquiry* 43, (2017): 102–107.

Todd, Zoe. "An Indigenous Feminist's Take on the Ontological Turn: 'Ontology' Is Just Another Word for Colonialism." *Journal of Historical Sociology* 29, no. 1 (2016): 4–22.

Tournier, V., C. M. Topham, A. Gilles, et al. "An Engineered PET Depolymerase to Break Down and Recycle Plastic Bottles." *Nature* 580, no. 7802 (2020): 216–219.

Tuck, Eve, and K. Wayne Yang. "Decolonization Is Not a Metaphor." *Decolonization: Indigeneity, Education & Society* 1, no. 1 (2012): 1–40.

Turner, Frederick J. "The Significance of the Frontier in American History." American Historical Association, January 1, 1893. Accessed October 24, 2024. https://www.historians.org/about-aha

-and-membership/aha-history-and-archives/historical-archives/the-significance-of-the-frontier-in
-american-history-(1893).

Turns, Anna. "Saving the Albatross: 'The War Is Against Plastic and They Are Casualties on the Frontline.'" *The Guardian*, March 12, 2018. Accessed May 26, 2024. https://www.theguardian .com/environment/2018/mar/12/albatross-film-dead-chicks-plastic-saving-birds.

Turns, Anna. "The Photo That Made the Plastics Crisis Personal." *BBC*, June 2 2023. Accessed October 24, 2024. https://www.bbc.com/future/article/20230531-the-photo-that-changed-the -worlds-response-to-the-plastics-crisis.

United Nations Environment Programme. "Historic Day in the Campaign to Beat Plastic Pollution: Nations Commit to Develop a Legally Binding Agreement." United Nations Environment Programme, March 2, 2022. Accessed March 10, 2023 https://www.unep.org/news-and-stories /press-release/historic-day-campaign-beat-plastic-pollution-nations-commit-develop.

Valente, Catherynne M. *The Past Is Red.* New York: Tom Doherty Associates, 2021.

van Dooren, Thom. *A World in a Shell: Snail Stories for a Time of Extinctions.* Cambridge, MA: The MIT Press, 2022.

Vehlken, Sebastian, Christina Vagt, and Wolf Kittler. "Introduction: Modeling the Pacific Ocean." *Media + Environment: Modeling the Pacific Ocean* 3, no. 2 (2021). https://doi.org/10.1525/

Voyles, Traci Brynne. *Wastelanding: Legacies of Uranium Mining in Navajo Country.* Minneapolis, MN; London: University of Minnesota Press, 2015.

Walford, Tone, and Lydia Gibson. "The Shapes of Things: Revising Geometries and Acknowledging Forms of/in/Through Territories, Terrains, and Technologies." Society for Social Studies of Science, 2023.

Walsh, Bryan. "The Truth about Plastic." *Time*, June 10, 2008. Accessed May 17, 2024. https:// time.com/archive/6596847/the-truth-about-plastic/.

Watts, Vanessa. "Indigenous Place-Thought Agency Amongst Humans and Non-Humans (First Woman and Sky Woman Go on a European World Tour!)." *Decolonization* 2, no. 1 (2013): 20–34.

Wehewehe Wikiwiki Hawaiin Language Dictionary. "Kamilo." Wehewehe Wikiwiki Hawaiin Language Dictionary. Accessed November 30, 2023. https://hilo.hawaii.edu/wehe/?q=kamilo#w2w2 -57658.

WHIM Architecture. "Recycled Island." Whim.nl, accessed August 21, 2014. Archived April 25, 2016 at https://web.archive.org/web/20160425130551/http://whim.nl/recycleisland.html.

White, Angel. "Oceanic 'Garbage Patch' Not Nearly as Big as Portrayed in Media." *Oregon State University: Newsroom*, January 04, 2011. Archived September 19, 2014 at https://web .archive.org/web/20140919020832/http://oregonstate.edu/ua/ncs/archives/2011/jan/oceanic -%E2%80%9Cgarbage-patch%E2%80%9D-not-nearly-big-portrayed-media.

Wong, May Ee. "Refiguring the 'Renewable Energy Island' as Future Environment." Society for the Social Study of Science, 2023.

Wu, Ming-Yi. *The Man with the Compound Eyes.* Translated by Darryl Sterk. Vintage Books, 2013.

Yang, Yu, Jun Yang, and Lei Jiang. "Comment on 'A Bacterium that Degrades and Assimilates Poly(Ethylene Terephthalate).'" *Science* 353, no. 6301 (2016): 759.

Yazzie, Melanie, and Cutcha Risling Baldy. "Introduction: Indigenous Peoples and the Politics of Water." *Decolonization: Indigeneity, Education & Society* 7, no. 1 (2018): 1–18.

Yoldas, Pinar. *Ecosystem of Excess.* 2014. https://pinaryoldas.info/Ecosystem-of-Excess-2014.

Yoshida, Shosuke, Kazumi Hiraga, Toshihiko Takehana, et al. "A Bacterium That Degrades and Assimilates Poly(Ethylene Terephthalate)." *Science* 351, no. 6278 (2016): 1196–1199.

Zettler, Erik R., Tracy J. Mincer, and Linda A. Amaral-Zettler. "Life in the 'Plastisphere': Microbial Communities on Plastic Marine Debris." *Environmental Science & Technology* 47, no. 13 (2013): 7137–7146.

Zheng Jiajia, and Suh Sangwon. "Strategies to Reduce the Global Carbon Footprint of Plastics." *Nature Climate Change* 9, no. 5 (2019): 374–378.

Zrimec, Jan, Mariia Kokina, Sara Jonasson, Francisco Zorrilla, and Aleksej Zelezniak. "Plastic-Degrading Potential Across the Global Microbiome Correlates with Recent Pollution Trends." *mBio* 12, no. 5 (2021). doi: https://doi.org/10.1101/2020.12.13.422558.

Zubiaurre, Maite. *Talking Trash: Cultural Uses of Waste.* Nashville, TN: Vanderbilt University Press, 2019.

Index

Note: Page numbers followed by *f* denote a figure.

ABC News, 88
ACC (American Chemistry Council), 13
Accountability, 17, 134, 135, 139–140
Ahupuaʻa (community), 43–44, 77
Alamitos Beach, 136
Alexander the Great, 42, 43
Algalita Marine Research and Education, 1–2, 8, 10, 25, 84–85, 110
American Chemistry Council (ACC), 13
American frontier, 7, 54–56, 58, 61, 62
Anthropocene epoch, 4, 18, 102, 104, 105, 109, 115, 118, 122
Anti-plastic sentiment, 14–15, 24, 106–109
  posters, 115–118, 116*f*, 119*f*
Aristotle, 45
Atomic Energy Commission, 57

Bacteria
  metabolizing plastics, 122–125, 126
  in the plastisphere, 103, 109, 120–121
Baldacchino, Godfrey, 51
Barad, Karen, 20, 22, 41, 73, 134
Barthes, Roland, 6
Biofilms, 102
Biopolitics, 105
Bios, vs. geos, 104–105
The Black Shoals (King), 135

"Blastic," 108
Blue Frontier project, 92
Borderwaters, 49
Boundaries. *See also* Water and land
  construction of, 22, 42, 53
  elemental philosophy and, 45–46, 50
  frontiers and, 5, 8, 54–55, 58, 122
  Indigenous peoples and, 134–136
  living/nonliving divide, 113, 114–115, 120
  in Plasticene epoch, 18–19
  traditional Hawaiian conceptions of, 43, 44
  Western concept of, 40, 42, 49, 103, 135–136, 140
#BreakFreeFromPlastic movement, 14
Burkhart, Brian, 7

California Coastal Commission, 115–118, 116*f*
Capitalism. *See also* Petrocapitalism
  colonialism and, 5, 17–18
  extraction of resources and, 43
  Liboiron on, 17
  ocean pollution and, 62–63
  plastics industry and, 5, 6
Carbon Imaginary, 104
Carson, Hank, 101–102, 109
ChatGPT, 69, 74

ChicoBag, 131
Cho, Seung Hee, 139
Cigarettes, 114, 115, 117–118, 119*f*
"Cig Egret" poster, 115, 116*f*, 117
Circular plastics economy, 124–125
Circulation, as method and site, 24–26
Clark, Eric, 51
Classification
    performativity, 49, 113
    sorting, 110–114, 118, 120
    taxonomy, 114, 118, 120
Clear Rivers project, 89–90
Climate change, 10, 16, 63, 75, 139
Coca-Cola, 13, 131
Colonialism
    boundary construction, 42
    capitalism and, 5, 17–18
    extraction of resources and, 17, 105
    fossil fuel extraction, 16
    green, 15–16
    Hawaii and, 135
    islands and, 51–54, 56–57, 92
    land, entitlement to, 105, 138
    "the line," 40, 41–43, 41*f*, 46, 51, 54, 62,
        104, 134–135, 138, 140–141
    nuclear, 57
    ocean pollution and, 62–63
    Pacific Ocean and, 50–57
    plastics industry and, 5
    pollution as, 17–18, 50, 62–63
    recycling and, 125
    Shotwell on, 23
    Turner and, 54
    waste, 57–58, 122
    wastelanding, 50, 52–53, 54–57
    water, 57
Colonization, biological, 103
Corpus Christi, Texas, 140
Critical theory, 122

da Cunha, Dilip, 40, 41–43, 51
Daily Mail, 82

Davis, Heather, 6, 17, 22, 104, 122, 127
DeLoughrey, Elizabeth, 51, 56–57
Diatoms, 103, 109
Diaz, Vicente, 135
Discover magazine, 69–70
Douglass, Frederick, 55
Duarte, Carlos, 95

Earth magazine, 83–84
Ebbesmeyer, Curtis, 10, 47–48
Eckert, Sandra, 125
EEZ (Exclusive economic zone), 49,
    130
"The 8th Continent" (Petráková), 61,
    91
Elemental philosophy, 45–46, 50, 52.
    See also Water and land
Endocrine disrupters, 12, 127
Entanglement, 19–24
    boundary construction, 22
    feminist new materialism and, 20–22
    fish crate as, 19–21, 20*f*
    harm, forms of, 21–22
    in marine ecology, 106–108
    of life and plastic, 114–115
    purity politics, 22–23
    saturation, 23
Enzymes, 124–125
Eriksen, Marcus, 66, 97, 131
Exclusive economic zone (EEZ), 49,
    130
Extraction of resources, 5, 14, 15
    capitalism and, 43
    colonialism and, 17, 105
    deep sea, 139
    fossil fuels, 16, 134
    harms of, 121
    Indigenous peoples, 44, 52
    Long Beach and, 140
    New Frontiers and, 56, 138
    petrocapitalism and, 138
    recycling and, 125

Feminist new materialism, 20–22, 72, 122

Feminist science and technology studies, 16–17, 74

Fiction, 59–61, 60*f*

Fish crate, entanglement and, 19–21, 20*f*

Floating architectures, 91–92. *See also* Recycled Island; Recycled Park; The 8th Continent; Seasteading Institute

Forever chemicals, 18. *See also* Persistent organic pollutants

Forms of concern, 63, 74, 94

Fortune magazine, 4, 4*f*

Fossil fuel extraction, 16, 134

Frontiers. *See also* Synthetic frontiers
  American, 7, 54–56, 58, 61, 62
  boundaries and, 5, 8, 54–55, 58, 122
  territorial power relations, 53–56

Fujikane, Candace, 44

Gabrys, Jennifer, 24

Ganges river, 42

Garbage patches. *See also* Great Pacific Garbage Patch; Trash island
  circulation as method and site, 24–26
  Ebbesmeyer and, 10
  gyres and, 11
  map of, 12*f*

Geos, vs. bios, 104–105

Ghost nets, 97–98, 98*f*

Gillis, John, 53

Global crises, responses to, 16

Goldstein, Miriam, 70–71, 72, 79, 84, 88

The Graduate, 15

Great Pacific (Harris and Morazo), 8, 9*f*, 59–61, 60*f*

Great Pacific Garbage Patch, 1–3, 11, 12*f*, 133
  images of, 71–72
  map of, 12*f*
  media descriptions of, 69–70, 71–72, 82–83
  model of, 48–50, 49*f*

predictions of, 46–48

public misconceptions of, 70–72, 79–83, 87

as synthetic frontier, 8, 57–61

Green colonialism, 15–16

Guam, 139

Guano Island Act, 54–55

Gyres, 11–12

Halobates, 121

Haraway, Donna, 20, 73

Harris, Joe, 9*f*, 59–61, 60*f*

Hauʻofa, Epili, 51

Hawaiʻi
  colonialism and, 135
  Kamilo Beach, 35–36, 37*f*, 38, 43
  traditional land divisions in, 43–45
  U.S. takeover of, 55

Hawaiʻi National Park, 55–56

Hegel, Georg, 7

Hegelian dialectic, 7

Helmreich, Stefan, 24, 94

Hester, Christina Dunbar, 139

HI-Zex Island, 92, 93*f*, 98*f*

Honolulu Commitment, 13–14

Ideonella sakaiensis, 123

Indian Ocean gyre, 11

Indigenous peoples, 134–135. *See also* Hawaiʻi; Kānaka Maoli
  boundaries and, 134–136
  displacement of, 55–56
  extraction of resources and, 44, 52
  land relations and, 135
  ontologies, 73, 105
  relational responsibility and, 43–44, 135–136
  wastelanding and, 52

Ingersoll, Karin Amimoto, 44

Ingraham, Jim, 47–48

International Coastal Cleanup, 115

The Invention of Rivers (da Cunha), 41

Islanding, 50, 51–53, 54–57
Islands
    colonialism and, 51–54, 56–57, 92
    ocean plastic pollution and, 73–74
    scholarship on, 51–52
    THUMS islands, 136–139, 137*f*, 140
    wastelanding and, 52

Jordan, Chris, 107, 121, 122
Jue, Melody, 23, 91

Kamilo Beach, 35–36, 37*f*, 38, 43
Kānaka Maoli (Native Hawaiian)
    abundance, 44–45
    boundaries, 43–45
    cartographies, 44
    genealogies, 45, 103–105
    responsibility, 44–45
    water, 43–45
Kennedy, John F., 56–57
Kewalo Basin Harbor, 40
King, Tiffany Lethabo, 135
Klaver, Irene, 51
Knoester, Ramon, 61, 75, 76–77, 88–91
Kuleana (responsibility), 44–45
Kumolipo, 103
Kurashima, Natalie, 44

Laist, David, 106
Land, entitlement to, 17, 105, 138
Land/water divide, 42–43, 46, 49–51, 56.
    See also Colonialism: "the line"; Water
    and land
Latour, Bruno, 23, 73, 74
Lattin, Gwen, 110
Lavoisier, Antoine, 45
Laysan albatross, 117, 121–122
Leslie, Ester, 7
Liboiron, Max, 17–18, 50, 121–122
Liliʻuokalani, Queen, 55
Lincoln, Noa, 44
Linton, Jamie, 45

Living/nonliving divide, 105–106, 111–112,
    114–115, 126
Living organisms
    living on ocean plastic pollution, 102–103,
        109–113, 111*f*
    ocean plastic pollution in, 113–120, 114*f*
    plastic species, 113–120, 116*f*, 119*f*
Long Beach, California, 136, 137*f*, 139, 140

Mah, Alice, 125
Manifest Destiny, 54
Marine debris, vs. plastic pollution, 13–14
Marshall Islands, 57
Masco, Joseph, 16
Materiality, 6, 17, 126
Matter and meaning, 20–23, 45–46, 72–74,
    93–94, 113. *See also* Entanglement;
    Forms of concern
Matters of care, 73, 74, 94
Matters of concern, 72–74, 94
Mawyer Alexander, 51
Meeting the Universe Halfway (Barad), 20–22,
    73
Meikle, Jeffrey, 5
Methodology, 24–26, 26*f*, 133
Micronesia, 57
Microorganisms, 79–80. *See also* Bacteria
Microplastics, 102–103
    anxiety about, 107–108
    bacteria and, 8, 120
    cigarettes and, 115
    definition of, 12
    images of, 84, 87
Midway Atoll, 55, 107
Military Waste (Reno), 52
Moore, Charles, 84
    Great Pacific Garbage Patch and, 8–10,
        46, 82
    on HI-ZEX Island, 92, 93*f*
    interview with, 71
    Plastic Ocean, 85
    plastic soup, 87–88

Morazzo, Martín, 9f, 59–61, 60f
Murphy, Michelle, 52

National parks model, 58
Nation of Hawai'i, 55–56
Netballs, 97–98, 98f
New Frontiers, 56–57, 138
New York Times, 70, 92
Nike, 47–48
"No Escape from Within," 108
"Non-Native Species of the California Coast,"
    117
North Atlantic Gyre, 11
North Pacific Gyre, 11, 36, 47
Nuclear colonialism, 16, 57

The Ocean Cleanup, 38–39, 39f, 49–50, 63,
    134
Ocean Conservancy, 115
Ocean current models, 47–48
Ocean plastic pollution, 8–14. *See also*
    Garbage patches; Great Pacific Garbage
    Patch; Trash island
    awareness of, 14–15, 25
    colonialism, capitalism, and racism, 62–63
    growth of, 15
    harm, images of, 107–108
    islands and, 73–74
    in living organisms, 113–120, 114f
    microorganisms and, 79–80
    Moore and, 8–10
    as new land, 59
    organisms living on, 102–103, 109–113,
        111f
    origins of, 10–11
    persistence of, 10, 14–19, 126–127,
        134–135
    as plastic soup, 85–87, 86f
    recycled island concept, 75–78, 76f,
        88–90
    separating samples, 109–113, 111f
    solutions for, 14, 16, 28, 59, 63, 124–126

techno fixes vs. systemic change, 38–40, 39f
trash island form, 94–95
Oceans, 51–53
Ocean Surface Current Simulation
    (OSCURS), 47
Oda, Kohei, 122–124
Ortho Plastic Novelties Inc., 4
OSCURS (Ocean Surface Current
    Simulation), 47

Pacific frontier, 27, 58
Pacific Ocean, 3
    colonialism and, 50–57
    garbage patches, 11, 12f, 24–26
    speculative landmasses, 72–73
"The Pacific Plastic Trash Island: A Growing
    Environmental Crisis" (ChatGPT), 69
Palmyrra, 55
Pearl Harbor, 140
Perkin, William Henry, 7
PET (Polyethylene terephthalate), 123
Petráková, Lenka, 61, 91
Petrocapitalism, 136–139, 137f
    plastivores and, 126–127, 133
    synthetic frontiers and, 139–141
Petrochemical production, 16
"Plague of Plastics" event, 115
Plankton, 110–111, 111f
Plastic, endurance of, 10, 14–19, 134–135
Plastic coastlines, 38–40, 39f, 46, 62–63, 136
Plasticene epoch, 18, 104
Plasticity, 6, 17
Plastic Kills! 108
Plastic Ocean (Moore), 85
Plastic pollution, 3. *See also* Ocean plastic
    pollution
    harm to humans, 108
    harm to the planet, 108–109
    images of, 136, 137f
    vs. marine debris, 13–14
    persistence of, 10, 14–19, 134–135
    photodegradation of, 12

Plastic production, 13–14
    attempts at reduction of, 115, 134
    in Hawaii, 62
    increases in, 15
    industry, 4–5
    Long Beach and, 140
    perpetuation of, 38
Plastic soup, 85–87, 86*f*
Plastic species, 113–120, 116*f*
    visual evolution of, 118–120, 119*f*
Plastic-to-plankton ratio, 110–111, 111*f*
Plastisphere, 138
    biosphere, compared to, 103–104
    definition of, 101–102
    earth's evolution, compared to, 103
    harms, mapping of, 121–122
    living/nonliving divide, 105–106
    plastic species, 113–120, 116*f*
Persistent organic pollutants (POPs), 18,
        126–127, 134, 136. *See also* Forever
        chemicals
Plastivores, 10, 28, 106, 121–125
    petrocapitalism and, 126–127, 133
Pollution as colonialism, 17–18, 50, 62–63
Polyethylene terephthalate (PET), 123
Polymers, 10, 102, 123
Port of Long Beach, 139. *See also* Long Beach,
        California
Povinelli, Elizabeth, 104–105, 109, 122
Power relations, 53–56, 94–95
Pravda, 71–72
Puig de la Bellacasa, Maria, 73, 74, 94
Purity politics, 22–23

Queer theory, 126

Racism, ocean pollution and, 62–63
"Rafting," 103
Recycled Island, 75–78, 76*f*, 88–90
Recycled Park, 90–91, 90*f*, 108
Recycling, 75, 76–78, 89, 123–126
Reno, Josh, 52, 53, 55

Re-turning, 3, 26, 133–134
RE:Villa, 89
Rivers, 42, 51
Ruiz, Rafico, 23

Sanford, Noni, 35–36
Sanford, Ron, 35–36
San Pedro Bay, 139
Science and technology studies (STS)
    classification, performativity of, 49
    feminist, 16–17, 74
    follow the actors, 24
Scripps Institute of Oceanography, 121
Sea Dragon expedition, 31–34, 40–41, 65–67,
        97–99, 129–132
SEA Lab, 110
Sea level rise, 59, 63, 75, 91, 134, 139
Sea of Japan, 47
Seasteading Institute, 91–92
Seattle Post Intelligencer, 82
Settler late liberalism, 104–105
Shih Ashanti, 55
Shotwell, Alexis, 22–23
Slat, Boyan, 39–40, 43, 59
Sommerville, Alice Te Punga, 135
South Atlantic gyre, 11
South Pacific gyre, 11
Speculative landmasses, 72–73
Spork Crab, 117
Starosielski, Nicole, 45
Steinberg, Philip, 53
STS. *See* Science and technology studies
SUPER (Survey of Underwater Plastic and
        Ecosystem Response), 80
Synthesis, 7
Synthetic, 6–7
"Synthetica: A New Continent of Plastics,"
        4–5, 4*f*
Synthetic chemists, 7
Synthetic circularity, 124–125
Synthetic frontiers
    accountability and, 139–140

definition of, 8, 58, 138
environmental crises and, 3
Great Pacific Garbage Patch as, 57–62
petrocapitalism and, 139–141
trash islands as, 8, 74, 75–78, 138
Synthetic Worlds (Leslie), 7

Teaiwa, Teresia, 51
Technological fix, 16, 28, 63, 91, 107
The Telegraph, 82
Terra Australia Incognita, 72
Terracentrism, 40, 42–43, 49, 61
Terraforming, 46, 139
THUMS islands, 136–139, 137f, 140
Time Magazine, 69
Todd, Zoe, 73, 127
Trash island, 1–3, 87–88
  accountability and, 135
  as misrepresentation, 1, 2, 24, 48, 70, 131,
    134
  as synthetic frontier, 8, 74, 75–78,
    138
  conceptions of, 94–95
  HI-Zex Island, 92, 93f, 98f
  power relations and, 94–95
Trask, Huanani-Kay, 135
Trichodesmium, 79–80
Turner, Frederick Jackson, 54, 56

United Nations Convention on the Law of
    the Seas (UNCLOS), 51
United Nations Environment Assembly, 15
United Nations Environment Program
    (UNEP), 12–13
United States
  Hawai'i and, 55
  New Frontiers, 56–57
  nuclear colonialism, 57
  Pacific Ocean and, 54–55
  sovereign ocean, 49
  wastelanding and islanding, 54–57
USS *Arizona*, 140

Vaughan, Mehana, 44–45
Voyles, Tracy, 52

Waste colonialism, 57–58, 122
Wastelanding, 50, 52–53, 54–57
Water and land. *See also* Boundaries; Land/
    water divide
  modern western conceptions of, 41–43,
    45–46, 50, 52, 58–59
  recycled island concept and, 76–77
  traditional Hawaiian conceptions of,
    43–45
Watts, Vanessa, 73
Western ontologies, 6, 19, 21, 41–43, 46, 51,
    103–105
WHIM architecture firm, 75, 77–78, 88
White, Angelicque, 79–83, 84
Wineera, Vernice, 135
Winfrey, Oprah, 83

Zubiaurre, Matie, 59

Publisher contact:
The MIT Press
Massachusetts Institute of Technology
77 Massachusetts Avenue, Cambridge, MA 02139
mitpress.mit.edu

EU Authorised Representative:
Easy Access System Europe, Mustamäe tee 50,
10621 Tallinn, Estonia
gpsr.requests@easproject.com

Printed by Integrated Books International,
United States of America